THE STAR GUIDE

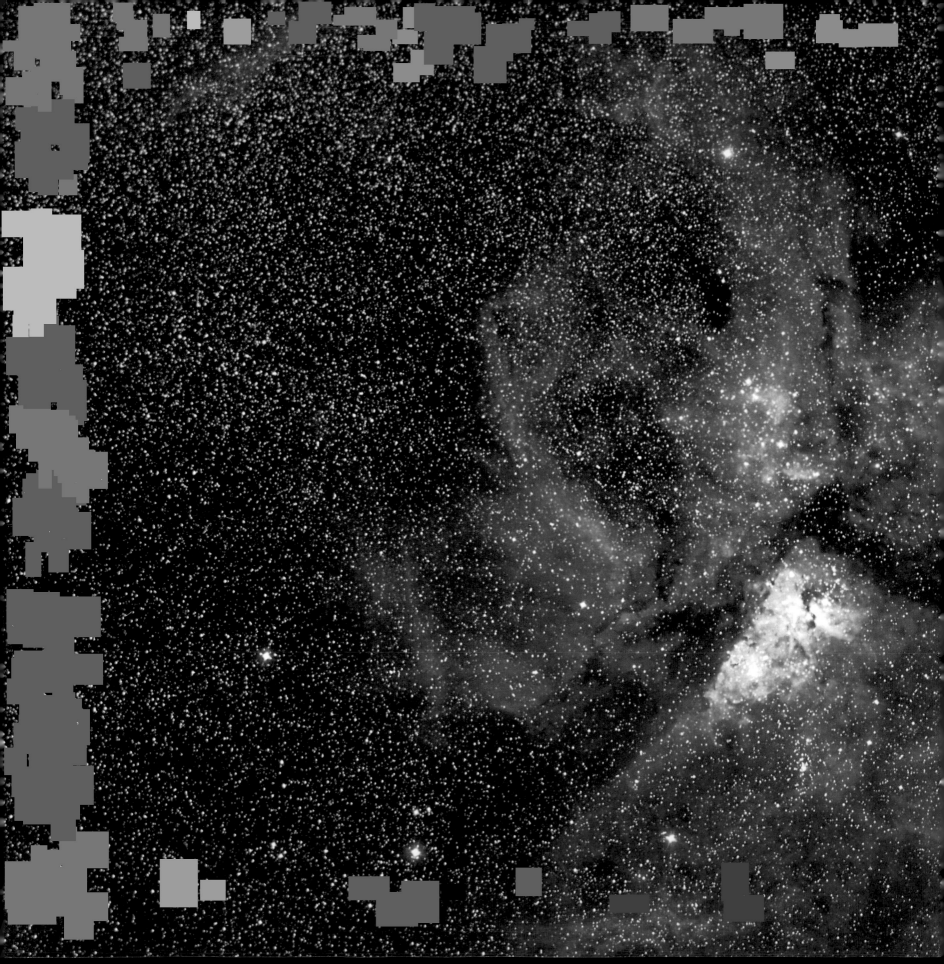

THE STAR GUIDE

LEARN HOW TO READ THE NIGHT SKY
STAR BY STAR

ROBIN KERROD

PRENTICE HALL GENERAL REFERENCE

NEW YORK LONDON TORONTO SYDNEY TOKYO SINGAPORE

A QUARTO BOOK

Copyright © 1993 by Quarto Inc.

Prentice Hall General Reference
15 Columbus Circle
New York, New York 10023

Library of Congress Cataloging-in-Publication Data

Kerrod, Robin
    The star guide: how to read the night sky star by star / Robin
  Kerrod. -- 1st Prentice Hall ed.
      p. cm.
    "A Quarto book" --T. p. verso.
    Includes index.
    ISBN 0-671-87467-5
    1. Stars--Observers' manuals. 2. Astronomy--Amateurs' manuals.
  I. Title.
  QB63.K45 1993
  523.8022'3--dc20   93-4818

Simultaneously published in Great Britain by Headline

This book was designed and produced by
Quarto Inc.
The Old Brewery
6 Blundell Street
London N7 9BH

Senior Art Editor Mark Stevens
Designer John Strange
Star map plots Richard Monkhouse and John Cox; David Kemp
(hemispheres 44-67)
Illustrations Janos Marffy; Ron Jobson (Moon maps 109-117)
Senior Editor Sally MacEachern
Copy Editor Eileen Cadman
Picture Research Manager Rebecca Horsewood
Art Director Moira Clinch
Publishing Director Janet Slingsby

With special thanks to Dr. Ron Maddison, Astronaut
Memorial Planetarium and Observatory, Florida

Typeset by CST, Eastbourne
Manufactured in Singapore by Eray Scan Pte. Ltd.
Printed in Hong Kong

10 9 8 7 6 5 4 3 2 1
First Prentice Hall Edition

CONTENTS

INTRODUCTION

The night sky has fascinated people since before the dawn of civilization. The first recorded writings in the Middle East, in cuneiform and hieroglyphics some 5,000 years ago, bear witness to a thorough knowledge of the stars and constellations. But, by and large, the heavens remained mysterious and awesome.

Not until the early seventeenth century did the night sky begin to give up some of its secrets. That was when Galileo trained his telescope on the heavens, and spied moons circling Jupiter and phases of Venus. It was the beginning of an age of astronomical enlightenment that in turn saw Kepler explain the motion of the planets, Newton elaborate the phenomenon of gravity, and Herschel discover Uranus.

But it was not until Bessel determined the distance to a nearby star, 61 Cygni, that the enormity of the stellar universe became apparent. Then, early this century, Hubble, using the newly completed 100-inch (254-centimeter) Hooker telescope, proved that spiral nebulae are separate island galaxies millions of light years away. The universe began to assume dimensions beyond the grasp of the human mind.

In recent years, radio astronomy and space astronomy by satellites and probes have brought to light an array of enigmatic bodies such as quasars and pulsars, bursters and blazars. It seems that the deeper we probe into the make-up of the universe, the more complex it becomes, confirming J.B.S. Haldane's remark: "The Universe is not only queerer than we imagine, it is queerer than we can imagine."

The star-studded firmament has lost some of its former mystery and, because of increasing light pollution, some of its former brilliance. But, when you are away from city lights, look up and enjoy the night sky with eyes, binoculars, and telescope. Travel in the mind's eye across interstellar space and ponder on the nature of the universe and on the meaning of it all.

And may you have clear skies.

STRUCTURE OF THE BOOK

Designed and structured to be user-friendly, this guide introduces the night sky with little preamble. Would-be observers do not have to wade through pages of astronomical information before feasting their eyes on the magnificent heavens. Yet extra information is always instantly available, via the novel icon-based cross-referencing system.

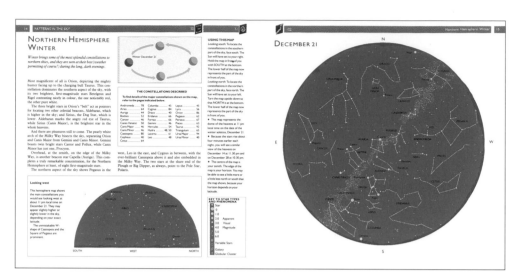

In the main, the book is organized on a spread-by-spread basis, with each spread covering a particular aspect of the night sky or presenting reference information on a specific topic. Each chapter of the book follows a similar pattern. After an opening spread, a two-page introduction precedes spreads on the featured subject. The chapter closes with a number of reference spreads.

▲ Seasonal star maps

The main maps in Chapter 1 present the first level of information about the night sky. They help observers familiarize themselves with the constellations that come into view season by season. Half-circle sketch maps are included to help interpretation, showing examples of sky views in a particular direction.

▶ Monthly star maps

The main maps in Chapter 2 provide more detailed information about the stars and constellations visible month by month throughout the year. They feature the brightest nebulae, clusters, and galaxies. Half-circle sketch maps are again included to help interpretation. They show the kinds of sky views northern and southern observers would see some time during the month.

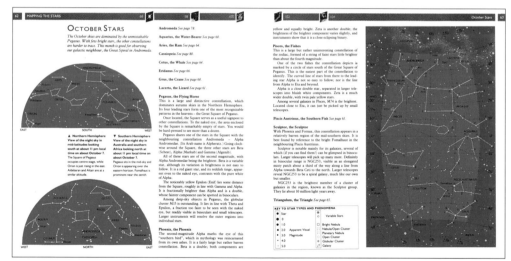

▶ Key constellations

The maps in Chapter 3 provide a third level of information about particular constellations, which by their richness merit more detailed investigation than usual by eye, binoculars, and telescope. They include the magnificent Sagittarius, which features dense star clouds, nebulae, and clusters.

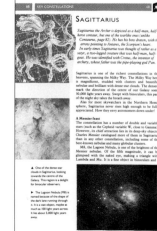

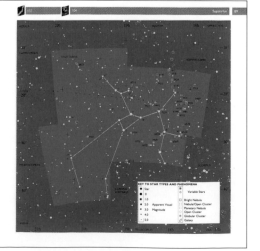

The maps

The first three chapters of the book focus on the stars and the constellations, and feature progressively more detailed star maps. The idea is that observers first become familiar with the shapes and locations of the constellations and then go on to investigate them in greater detail.

The main maps in Chapter 1 present the heavens season by season. They show the constellations that appear in the night sky in winter, spring, summer, and fall. Maps are included for both the Northern and the Southern Hemispheres. Half-circle sky views show the constellations visible looking west.

The main maps in Chapter 2 feature the stars and constellations prominent in the sky in particular months. They also include the brightest star clusters, nebulae, and galaxies. Half-circle sky views help northern and southern observers orientate themselves. The maps in Chapter 3 present more detailed information on particular outstanding constellations.

The maps in Chapter 4 feature the Moon. They cover the near side – the one that always faces us – in four quadrants. The seas, the main craters, mountain ranges, and other features of the lunar surface are marked.

Cross-referencing

The main text spreads in the book are concerned primarily with observation and practical astronomy in general. They can be read and appreciated quite independently of the reference spreads that appear at the end of each chapter. But if you want further information, you can turn to the reference spreads, using the icons.

If, for example, you are reading about the Orion Nebula and want to know more about nebulae, look for the "Nebula" icon in the band and note the number beside it (100). Turn to page 100 and you will find detailed information on nebulae. If, at any time, you want to refresh your memory on common astronomical terms, turn to the Glossary on pages 156 and 157.

▶ **Moon maps**

The maps in Chapter 4 feature our nearest neighbor in space, the Moon. There are four maps, each covering one quadrant. The maps are printed with north at the top, and show the Moon as we see it with the naked eye and through binoculars. The maria (seas) and most prominent craters are marked.

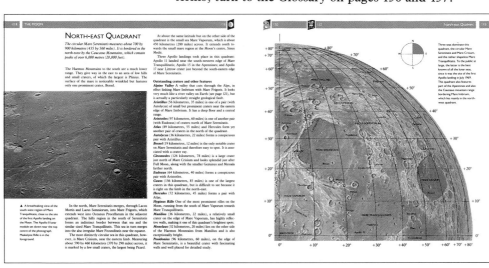

The Lunar Surface

Outer Galaxies

Star Anatomy

Star Clusters

Moon Movements

The Milky Way

Nebulae

The Apollo Explorations

◀ **Reference spreads**

This is an example of a reference spread. Reference spreads appear at the end of each chapter and provide background information on a variety of topics.

▶ **Cross-reference icons**

Examples of the icons that appear, together with page numbers, in the band running along the top of ordinary text spreads. They refer to reference spreads focused on particular subjects.

◄ Fanciful figures that represent some constellations of the Northern Hemisphere. The zodiacal constellations Virgo, Libra, Scorpius, and Sagittarius feature prominently in the picture, although Virgo and Libra are far from prominent in the heavens.

PATTERNS IN THE SKY

On a clear moonless night, the heavens present a dazzling spectacle. Myriads of twinkling stars sparkle like jewels against the velvety backcloth of space.

Even a quick glance at the night sky reveals that the stars are not all the same. Most obviously, they vary in brightness; they also show subtle variations in color.

If we remain stargazing for some time, we can easily learn to recognize the patterns the bright stars make – patterns we call the constellations. Familiarizing ourselves with the constellations is the first step in stargazing, for they act as celestial signposts to help us find our way around the night sky.

Familiarizing ourselves with the constellations is the object of this chapter. We find out how and why they appear in, and disappear from, the night sky throughout the year, as one season gives way to the next.

The star maps represent the appearance of the late evening sky on significant seasonal dates – midwinter and midsummer and the equinoxes. Maps are included for both the Northern and the Southern Hemispheres.

ANCIENT PATTERNS

The patterns of stars we see in the sky today will not change perceptibly in our lifetime. Indeed, they have changed scarcely at all since the time of the first great civilizations some 5,000 years ago.

In those days, priest-astronomers observed the rhythms of the heavens to establish calendars, which could bring a semblance of order into the day-to-day lives of an expanding population in an increasingly complex society. They also looked to the heavens for signs that would indicate the will of their gods, whom they believed ruled human lives. Their "reading the stars" led to the pseudo-science of astrology, which is still with us. Astrologers reckon that the relative positions of the heavenly bodies control human destiny.

▼ One of the most distinctive star patterns in the heavens is the constellation of Orion. It sits on the celestial equator and is therefore visible to star-gazers in both the Northern and the Southern Hemispheres. The sketch shows the way we join up the brightest stars to form the main constellation outline.

The celestial sphere

The records kept by those learned ancients show graph-ically that the constellations they were familiar with are virtually identical to those we see today. It seems, there-fore, as if their concept of the heavens is valid. They considered the stars to be fixed in position on the inside of a great celestial sphere enveloping the Earth.

We now know that there is no great celestial sphere, but, nevertheless, the concept is useful in astronomy, for it provides a convenient way of pinpointing the positions of the stars in the sky. (Turn to pages 34–5 to see how astronomers use it.)

The whirling heavens

If you remain stargazing for any length of time, you notice that the stars move across the sky. They do not move relative to one another, but as a whole. In most parts of the world, if you follow a particular star through-out the night, you find that it first appears over the eastern horizon as darkness falls. Then it arcs up through the

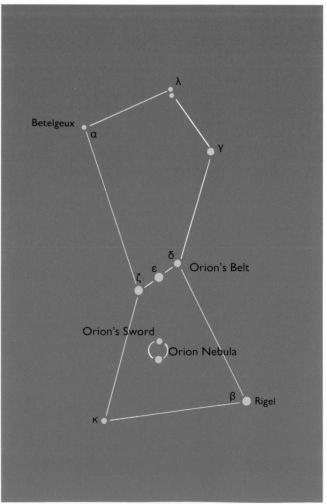

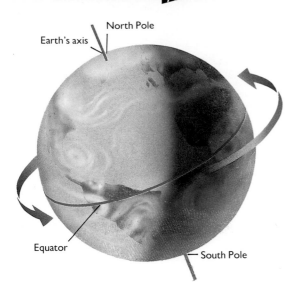

North Pole

Earth's axis

Equator

South Pole

▲ The reason why the stars wheel overhead: the Earth spins on its axis, making one revolution a day. We show the axis tilted to reflect the fact that it is tilted in relation to the direction it travels through space.

sky, reaching its highest point in the south. It then arcs down again, eventually disappearing over the western horizon when daylight returns. Most of the stars arc through the night sky in a similar way, traveling from east to west, as the Sun does.

The celestial sphere, then, appears to be rotating from east to west around the Earth. But actually, the reverse is true. It is the Earth that is moving relative to the stars. It spins around once a day on its axis, an imaginary line passing through the North and South Poles.

It is because of the rotation of the Earth that we must specify on a star map a particular time of night when the stars will appear in the sky in the positions shown on the map. At other times, the stars will appear in different positions in the sky.

Changing with the seasons

If you look at the dome of the heavens over a period of months, you will notice that some constellations are always present. Others, however, come and go. This is a consequence of the Earth's other motion in space – it travels in orbit around the Sun once a year (see diagram, right).

Again, from our Earth-bound viewpoint, it is the Sun that appears to move. It appears to travel around the celestial sphere against the background of stars once a year. At any time, we will not be able to see the constellations that lie in the same direction as the Sun, because the Sun's light will blot them out – it will be daylight. As the Sun proceeds in its path around the celestial sphere during the year (the ecliptic), different constellations will be blotted out in turn while others return to dark skies.

The seasonal star maps

The maps that follow show views of the night sky as it appears in late evening in mid-latitudes in the Northern and Southern Hemispheres in midwinter and midsummer and at the time of the March and September equinoxes.

The circular maps represent the dome of the heavens above your head. Only the main constellations are shown. Which of the constellations you will see at any instant will depend on which direction you look in. As an example, we include on each spread a semicircular map that shows the constellations you would see looking west. West was chosen because it is an easy direction to find, being the direction in which the Sun sets.

Using these maps will help you identify the constellations and get your celestial bearings as it were, season by season. The maps most nearly represent the night sky only at the specific time and on the specific dates chosen, but for the purposes of constellation recognition and general orientation, they are usable for much longer.

▼ Key points in the annual orbit of the Earth around the Sun. The tilt of the Earth's axis always remains the same, which means that each hemisphere is angled more toward the Sun at some times of the year than at others. This causes a difference in temperature throughout the year and brings about the seasons. The dates are of midsummer, midwinter, and the equinoxes. The seasons are opposite in the Northern and Southern Hemispheres.

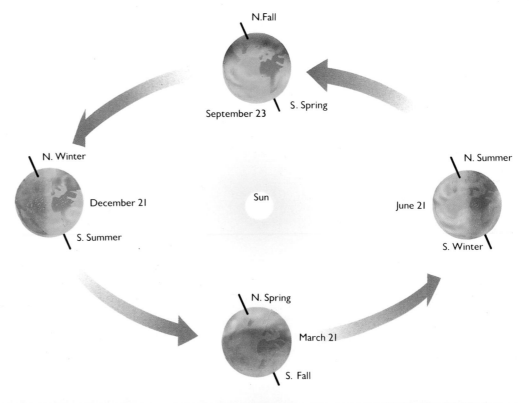

N.Fall

September 23   S. Spring

N. Winter   N. Summer

December 21   Sun   June 21

S. Summer   S. Winter

N. Spring

March 21

S. Fall

NORTHERN HEMISPHERE WINTER

Winter brings some of the most splendid constellations to northern skies, and they are seen at their best (weather permitting, of course!) during the long, dark evenings.

Most magnificent of all is Orion, depicting the mighty hunter facing up to the charging bull Taurus. This constellation dominates the southern aspect of the sky, with its two brightest, first-magnitude stars Betelgeux and Rigel contrasting nicely in color, the one noticeably red, the other pure white.

The three bright stars in Orion's "belt" act as pointers for locating two other celestial beacons, Aldebaran, which is higher in the sky; and Sirius, the Dog Star, which is lower. Aldebaran marks the angry red eye of Taurus, while Sirius (Canis Major), is the brightest star in the whole heavens.

And there are pleasures still to come. The pearly white arch of the Milky Way bisects the sky, separating Orion and Canis Major from Gemini and Canis Minor. Gemini boasts twin bright stars Castor and Pollux, while Canis Minor has just one, Procyon.

Overhead, at the zenith, on the edge of the Milky Way, is another beacon star Capella (Auriga). This completes a truly remarkable concentration, for the Northern Hemisphere at least, of eight first-magnitude stars.

The northern aspect of the sky shows Pegasus in the west, Leo in the east, and Cygnus in between, with the ever-brilliant Cassiopeia above it and also embedded in the Milky Way. The two stars at the cup end of the Big Dipper, as always, point to the Pole Star, Polaris.

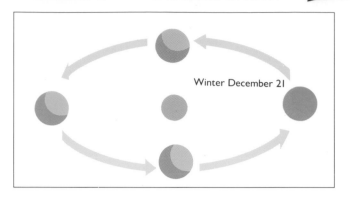

Winter December 21

THE CONSTELLATIONS DESCRIBED

To find details of the major constellations shown on the map, refer to the pages indicated below.

Andromeda 78	Columba 45	Lepus 45
Aries 64	Cygnus 84	Lyra 56
Auriga 44	Draco 40	Orion 86
Boötes 52	Eridanus 66	Pegasus 62
Cancer 46	Fornax 66	Perseus 67
Canes Venatici ... 50	Gemini 46	Pisces 63
Canis Major 44	Hercules 54	Taurus 92
Canis Minor 46	Hydra 48, 50	Triangulum 65
Cassiopeia 80	Lacerta 61	Ursa Major 94
Cepheus 40	Leo 48	Ursa Minor 40
Cetus 64		

USING THIS MAP

- This map represents the dome of the heavens at 11 p.m. local time on the date of the winter solstice, December 21.
- Because the stars rise about four minutes earlier each night, you will see a similar view of the heavens on December 14 at 11.30 p.m. and on December 28 at 10.30 p.m.
- The center of the map is your zenith. The edge of the map is your horizon. You may be able to see a little more or a little less north or south than the map shows, because your horizon depends on your latitude.

Looking south: To locate the constellations in the southern part of the sky, face south. The Sun will have set on your right. Hold the map in front of you with SOUTH at the bottom. The lower half of the map now represents the part of the sky in front of you.

Looking north: To locate the constellations in the northern part of the sky, face north. The Sun will have set on your left. Turn the map upside-down so that NORTH is at the bottom. The lower half of the map now represents the part of the sky in front of you.

Looking west

This hemisphere map shows the main constellations you would see looking west at about 11 p.m. local time on December 21. They may appear slightly higher or slightly lower in the sky, depending on your exact latitude.

The unmistakable W-shape of Cassiopeia and the Square of Pegasus are prominent.

Aldebaran

PERSEUS

TAURUS

Pleiades

ARIES

CASSIOPEIA

CEPHEUS

PEGASUS

Deneb

CETUS

CYGNUS

SOUTH   WEST   NORTH

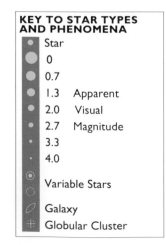

KEY TO STAR TYPES AND PHENOMENA

Star	
0	
0.7	
1.3	Apparent
2.0	Visual
2.7	Magnitude
3.3	
4.0	
	Variable Stars
	Galaxy
	Globular Cluster

DECEMBER 21

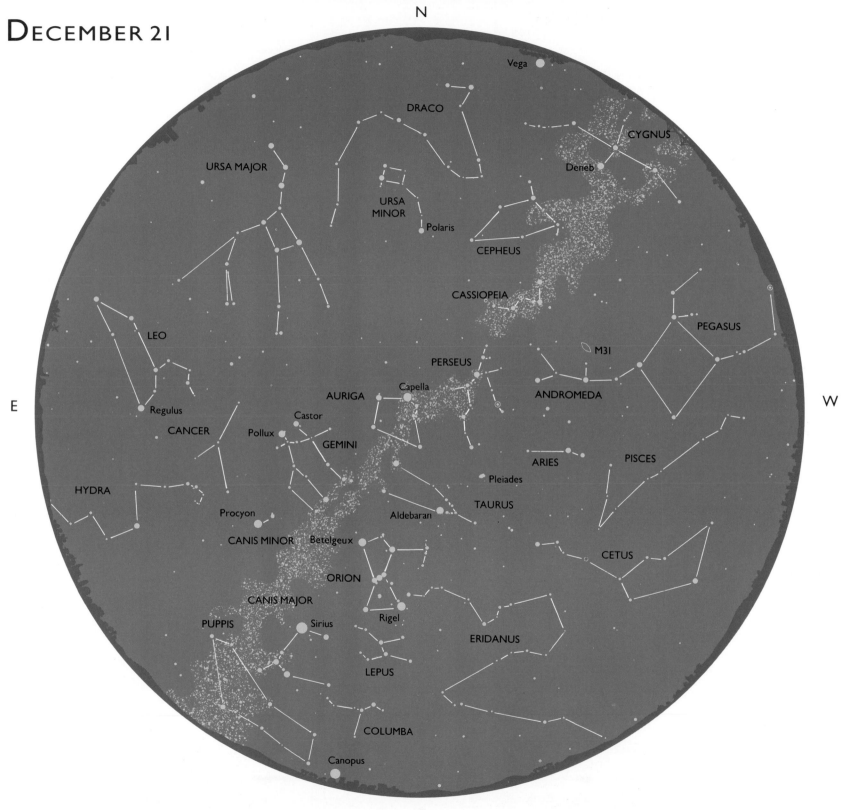

N

E

W

S

Vega

DRACO

CYGNUS

Deneb

URSA MAJOR

URSA MINOR

Polaris

CEPHEUS

CASSIOPEIA

PEGASUS

LEO

PERSEUS

M31

ANDROMEDA

Capella

AURIGA

Regulus

Castor

CANCER

Pollux

GEMINI

ARIES

PISCES

Pleiades

HYDRA

TAURUS

Procyon

Aldebaran

CANIS MINOR

Betelgeux

CETUS

ORION

CANIS MAJOR

ERIDANUS

PUPPIS

Sirius

Rigel

LEPUS

COLUMBA

Canopus

NORTHERN HEMISPHERE SPRING

With the coming of spring, Orion is setting by the late evening, although there is still plenty of time after sunset to view it.

The southern aspect of the sky in late evening belongs to Leo, whose distinctive back-to-front question-mark star pattern (the Sickle) traces the Lion's head and mane.

Arcing southwest toward Orion, the twins Castor and Pollux (Gemini) are still prominent. Between Leo and Gemini, Cancer contains one of the top three star clusters in the northern heavens, Praesepe (Beehive).

Castor and Pollux are approximate pointers, aiming southwest to the brighter Procyon (Canis Minor). Farther south, Canis Major and its prime star Sirius are close to setting.

In the east, the only two really bright stars are Spica (Virgo) and Arcturus (Boötes). Arcturus is a warm orange color and the brightest star in the Northern Hemisphere.

The northern aspect of the sky reveals the Big Dipper high up. Cassiopeia is fairly low in the northwest, pointing to the brilliant Capella (Auriga). Seemingly linked with Auriga is Taurus, whose chief star is Aldebaran. Its reddish hue contrasts noticeably with Capella's pure bluishwhite. Taurus's other main naked-eye attractions are its two prominent star clusters, the Hyades, surrounding Aldebaran, and the Pleiades, or Seven Sisters.

In the east Deneb (Cygnus) and Vega (Lyra) appear close to the horizon. The distinctive crescent of the Northern Crown (Corona Borealis) appears higher up.

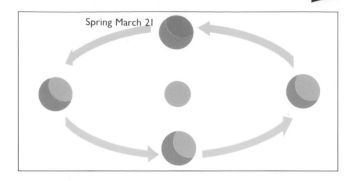

Spring March 21

THE CONSTELLATIONS DESCRIBED

To find details of the major constellations shown on the map, refer to the pages indicated below.

Andromeda 78	Corona	Monoceros 47
Aries 64	Borealis 54	Orion 86
Auriga 44	Cygnus 84	Perseus 67
Boötes 52	Draco 40	Pisces 63
Cancer 46	Eridanus 66	Puppis 47
Canes Venatici 50	Fornax 66	Serpens Caput 55
Canis Major 44	Gemini 46	Taurus 92
Canis Minor 46	Hercules 54	Triangulum 65
Cassiopeia 80	Hydra 48, 50	Ursa Major 94
Cepheus 40	Lacerta 61	Ursa Minor 40
Columba 45	Leo 48	Virgo 96
Coma Berenices. 52	Lyra 56	

USING THIS MAP

● This map represents the dome of the heavens at 11 p.m. local time on the date of the spring, or vernal equinox, March 21.

● Because the stars rise about four minutes earlier each night, you will see a similar view of the heavens on March 14 at 11.30 p.m. and on March 28 at 10.30 p.m.

● The center of the map is your zenith. The edge of the map is your horizon. You may be able to see a little more or a little less north or south than the map shows, because your horizon depends on your latitude.

Looking south: To locate the constellations in the southern part of the sky, face south. The Sun will have set on your right. Hold the map in front of you with SOUTH at the bottom. The lower half of the map now represents the part of the sky in front of you.

Looking north: To locate the constellations in the northern part of the sky, face north. The Sun will have set on your left. Turn the map upside-down so that NORTH is at the bottom. The lower half of the map now represents the part of the sky in front of you.

Looking west

This hemisphere map shows the main constellations you would see looking west at about 11 p.m. local time on March 21. They may appear slightly higher or slightly lower in the sky, depending on your exact latitude.

This view boasts a collection of bright stars, including Sirius, Betelgeux, and Capella.

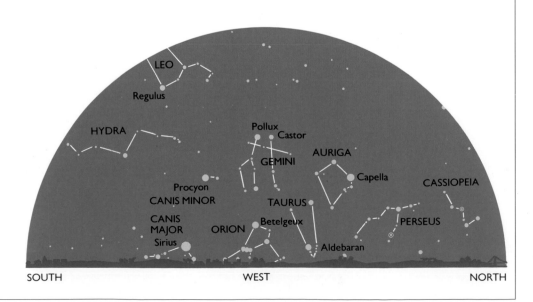

LEO
Regulus
HYDRA
Pollux
Castor
AURIGA
GEMINI
Capella
CASSIOPEIA
Procyon
CANIS MINOR
TAURUS
PERSEUS
CANIS MAJOR
ORION
Betelgeux
Sirius
Aldebaran

SOUTH   WEST   NORTH

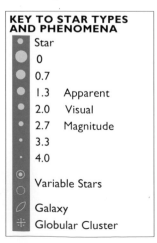

KEY TO STAR TYPES AND PHENOMENA

●	Star	
●	0	
●	0.7	
●	1.3	Apparent
●	2.0	Visual
●	2.7	Magnitude
●	3.3	
·	4.0	
⊙	Variable Stars	
○		
⬭	Galaxy	
✷	Globular Cluster	

March 21

N

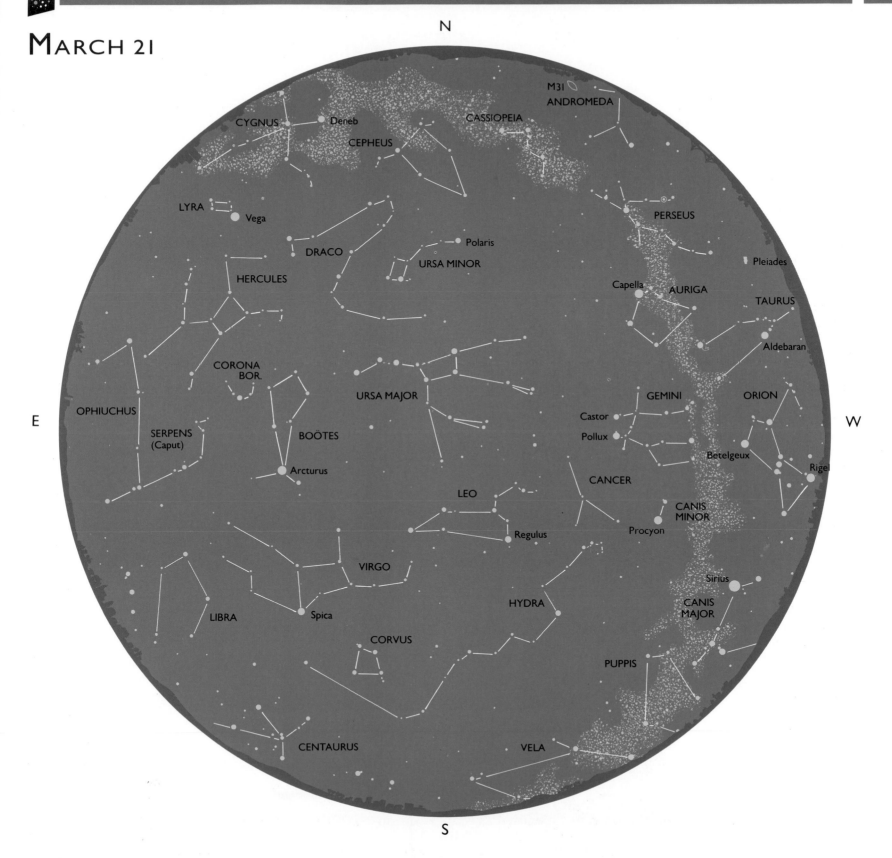

E

W

S

Northern Hemisphere Summer

The warmer nights make summer stargazing a pleasure, although their drawback is that the skies are never really dark.

High in the sky at this time of the summer solstice are three first-magnitude stars, Deneb (Cygnus), Vega (Lyra), and Altair (Aquila). They form the celebrated Summer Triangle. It overlays a rich region of the Milky Way, although it is difficult to appreciate it fully because of the light skies.

Deneb, in the tail of the flying Swan, is the dimmest of the summer trio, at least to the eye. In reality, it is thousands of times more luminous, appearing dimmer only because it is more than 60 and 100 times farther away respectively than Vega and Altair.

Deneb shines high in the northeast, while above the eastern horizon is Andromeda and part of the Square of Pegasus, which will become prominent in fall skies. Leo, meanwhile, is about to set in the far west.

The other two stars of the Summer Triangle, Vega and Altair, appear in southeastern skies, the one high up, the other midway to the horizon. Rising just above the horizon here are portions of two of the splendid Southern Hemisphere constellations that northern astronomers drool over. They are Sagittarius and, nearly due south, Scorpius, whose pulsating red supergiant star Antares

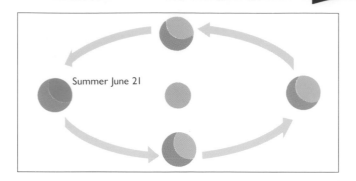

Summer June 21

THE CONSTELLATIONS DESCRIBED

To find details of the major constellations shown on the map, refer to the pages indicated below.

Andromeda 78	Delphinus 58	Pegasus 62
Aquarius 60	Draco 40	Perseus 67
Aquila 58	Eridanus 66	Puppis 47
Auriga 44	Fornax 66	Sagittarius 88
Boötes 52	Gemini 46	Scorpius 90
Cancer 46	Hercules 54	Scutum 57
Canes Venatici 50	Lacerta 61	Serpens Caput ... 55
Cassiopeia 80	Leo 48	Serpens Cauda ... 57
Cepheus 40	Libra 53	Ursa Major 94
Columba 45	Lynx 49	Ursa Minor 40
Coma Berenices 52	Lyra 56	Virgo 96
Corona Borealis 54	Monoceros 47	Vulpecula 59
Cygnus 84	Ophiuchus 56	

("Rival of Mars") marks the Scorpion's heart.

Antares forms one corner of a more widely separated "summer triangle" with Spica (Virgo), at about the same elevation, and Arcturus (Boötes) farther north.

USING THIS MAP

● This map represents the dome of the heavens at 11 p.m. local time on the date of the summer solstice, June 21.
● Because the stars rise about four minutes earlier each night, you will see a similar view of the heavens on June 14 at 11.30 p.m. and on June 28 at 10.30 p.m.
● The center of the map is your zenith. The edge of the map is your horizon. You may be able to see a little more or a little less north or south than the map shows, because your horizon depends on your latitude.

Looking south: To locate the constellations in the southern part of the sky, face south. The Sun will have set on your right. Hold the map in front of you with SOUTH at the bottom. The lower half of the map now represents the part of the sky in front of you.

Looking north: To locate the constellations in the northern part of the sky, face north. The Sun will have set on your left. Turn the map upside-down so that NORTH is at the bottom. The lower half of the map now represents the part of the sky in front of you.

Looking west

This hemisphere map shows the main constellations you would see looking west at about 11 p.m. local time on June 21. They may appear slightly higher or slightly lower in the sky, depending on your exact latitude.

Arcturus straight ahead forms a prominent triangle with Antares and Spica to the south.

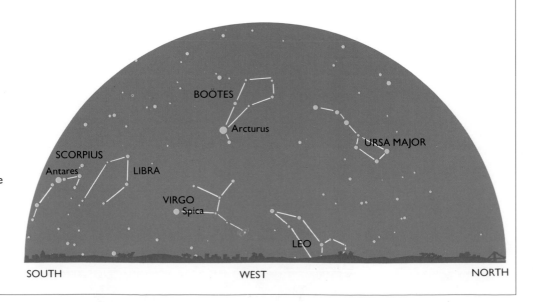

SOUTH   WEST   NORTH

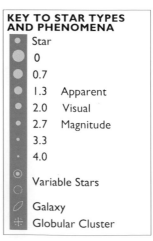

KEY TO STAR TYPES AND PHENOMENA

Star	
0	
0.7	
1.3	Apparent
2.0	Visual
2.7	Magnitude
3.3	
4.0	
Variable Stars	
Galaxy	
Globular Cluster	

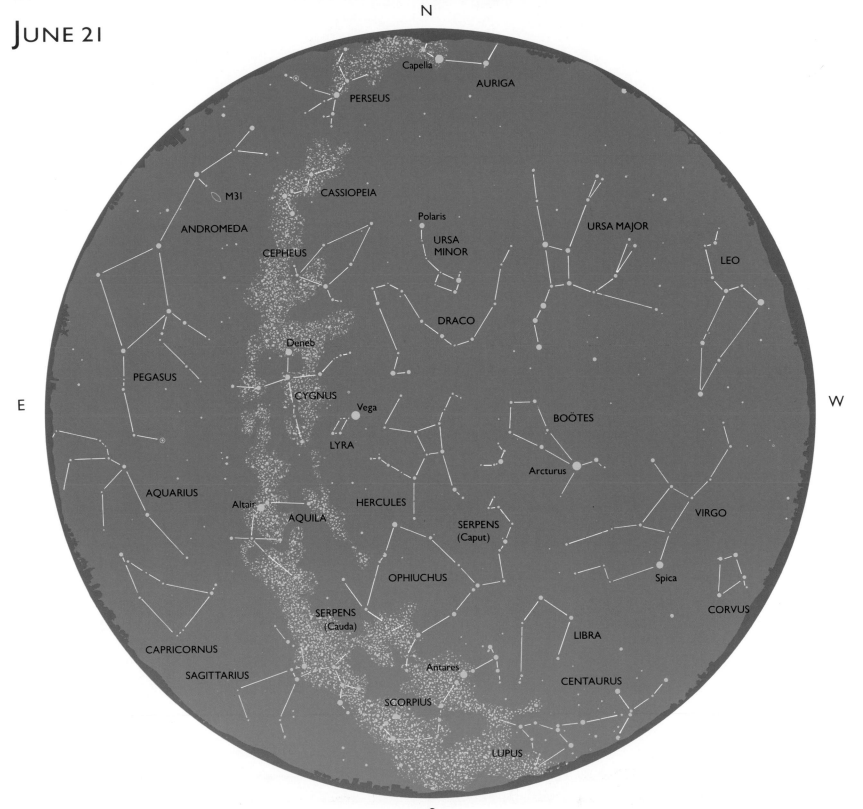

JUNE 21

N

Capella

AURIGA

PERSEUS

CASSIOPEIA

M31

ANDROMEDA

CEPHEUS

Polaris

URSA MINOR

URSA MAJOR

LEO

DRACO

Deneb

PEGASUS

CYGNUS

Vega

E   W

BOÖTES

LYRA

Arcturus

AQUARIUS

HERCULES

Altair

AQUILA

SERPENS (Caput)

VIRGO

OPHIUCHUS

Spica

SERPENS (Cauda)

CORVUS

LIBRA

CAPRICORNUS

Antares

CENTAURUS

SAGITTARIUS

SCORPIUS

LUPUS

S

NORTHERN HEMISPHERE FALL

With evening skies darkening perceptibly, the Milky Way can now be better appreciated as it cuts across the sky from northeast to southwest.

Looking south, this silvery arch carries with it Cygnus high overhead, and Aquila lower down. The Summer Triangle, formed by their two main stars, Deneb and Altair, together with Vega (Lyra), is still conspicuous, though not for long.

The late evening sky is now graced by Pegasus with its unmistakable Square. The two stars in the westernmost side of the Square act as pointers to Fomalhaut (Piscis Austrinus), just visible above the southern horizon.

The line of stars running east from the northern side of the Square belong to Andromeda. This constellation's main claim to fame is the fuzzy patch of light we can see just to the north of the line of stars. This patch, once thought to be a nebula in our own Galaxy, turns out to be a quite separate galaxy, much bigger than our own, but incredibly distant (some 13.5 million million million miles, 22 million million million kilometers!).

Looking north, the Big Dipper is near its lowest point, while on the other side of Polaris, Cassiopeia is near its highest. While much of Boötes is visible in the northwest, its bright star Arcturus is below the horizon. Close by is the distinctive curve of the Northern Crown (Corona Borealis).

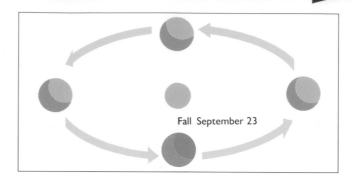

Fall September 23

THE CONSTELLATIONS DESCRIBED

To find details of the major constellations shown on the map, refer to the pages indicated below.

Andromeda 78	Corona Borealis . 54	Piscis Austrinus 61
Aquarius 60	Cygnus 84	Puppis 47
Aquila 58	Delphinus 58	Sagittarius 88
Aries 64	Draco 40	Sculptor 63
Auriga 44	Hercules 54	Scutum 57
Boötes 52	Hydra 48, 50	Serpens Caput 55
Cancer 46	Lacerta 61	Serpens Cauda ... 57
Canes Venatici 50	Lynx 49	Taurus 92
Canis Minor 46	Lyra 56	Triangulum 65
Capricornus 58	Ophiuchus 56	Ursa Major 94
Cassiopeia 80	Pegasus 62	Ursa Minor 40
Cepheus 40	Perseus 67	Vulpecula 59
Cetus 64	Pisces 63	
Columba 45		

There is more interest, however, in the east where Auriga and Taurus have now risen, with their two brilliant main stars, Capella and Aldebaran, set for prominence in winter skies.

USING THIS MAP

● This map represents the dome of the heavens at 11 p.m. local time on the date of the Fall equinox, September 23.
● Because the stars rise about four minutes earlier each night, you will see a similar view of the heavens on September 16 at 11.30 p.m. and on September 30 at 10.30 p.m.
● The center of the map is your zenith. The edge of the map is your horizon. You may be able to see a little more or a little less north or south than the map shows, because your horizon depends on your latitude.

Looking north: To locate the constellations in the northern part of the sky, face north. The Sun will have set to your left. Hold the map in front of you with NORTH at the bottom. The lower half of the map now represents the part of the sky in front of you.

Looking south: To locate the constellations in the southern part of the sky, face south. The Sun will have set to your right. Turn the map upside-down so that SOUTH is at the bottom. The lower half of the map now represents the part of the sky in front of you.

Looking west

This hemisphere map shows the main constellations you would see looking west at about 11 p.m. local time on September 23. They may appear slightly higher or slightly lower in the sky, depending on your exact latitude.

The three stars of the Summer Triangle – Altair, Deneb, and Vega – are straight ahead.

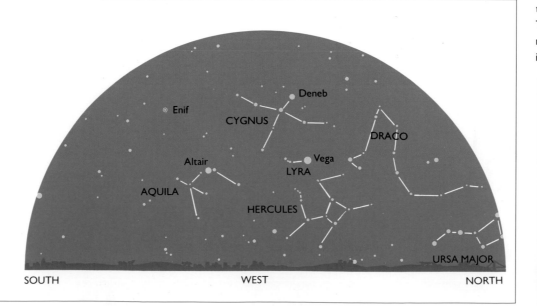

SOUTH   WEST   NORTH

KEY TO STAR TYPES AND PHENOMENA

Star	
0	
0.7	
1.3	Apparent
2.0	Visual
2.7	Magnitude
3.3	
4.0	
Variable Stars	
Galaxy	
Globular Cluster	

September 23

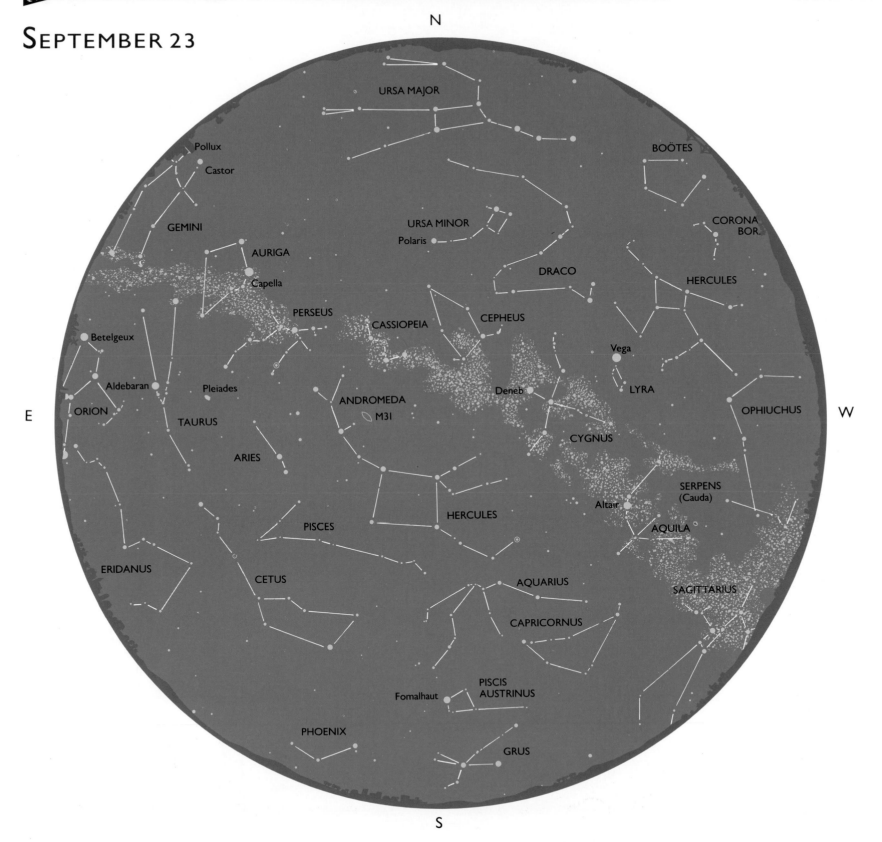

N

Pollux
Castor
GEMINI
AURIGA
Capella
PERSEUS
URSA MAJOR
BOÖTES
CORONA BOR.
URSA MINOR
Polaris
DRACO
HERCULES
CASSIOPEIA
CEPHEUS
Vega
Betelgeux
LYRA
Deneb
Aldebaran
ANDROMEDA
M31
CYGNUS
OPHIUCHUS
ORION
TAURUS
Pleiades
SERPENS (Cauda)
ARIES
Altair
HERCULES
AQUILA
PISCES
SAGITTARIUS
ERIDANUS
CETUS
AQUARIUS
CAPRICORNUS
PISCIS AUSTRINUS
Fomalhaut
PHOENIX
GRUS

E   W

S

SOUTHERN HEMISPHERE SUMMER

While astronomers in the Northern Hemisphere shiver in midwinter, their colleagues "down under" in the Southern Hemisphere enjoy the balmy nights of mid-summer.

As in the Northern Hemisphere, Orion is a dominant constellation, seen by southern astronomers in the northern aspect of the sky. They see it with white Rigel in the upper left of the main pattern, and the reddish Betelgeux in the bottom right.

Following a line through the three stars of Orion's "belt" leads higher up to the brightest star in the heavens, Sirius (Canis Major) and lower down to the reddish Aldebaran (Taurus). Continuing the same line through Aldebaran leads to the Pleiades, or Seven Sisters, star cluster. And near the horizon beneath Aldebaran is the brighter Capella (Auriga).

East of Capella are the twins, Castor and Pollux (Gemini), with the brilliant Procyon (Canis Minor) higher up. The brilliant stars in this corner of the sky together form a prominent hexagonal pattern.

The southern aspect of the sky, as always in the Southern Hemisphere, is spectacular. The second brightest star in the heavens, Canopus (Carina), sits high in the sky, as does Achernar (Eridanus) farther west. Farther west still and lower is Fomalhaut (Piscis Austrinus).

The Southern Cross (Crux) and its two pointers, Alpha and Beta Centauri (Centaurus), are stunning as always, and seen relatively close to the horizon. The fuzzy patches we see southwest of Canopus are neighboring galaxies, the Large (Doradus) and Small Magellanic Cloud (Tucana).

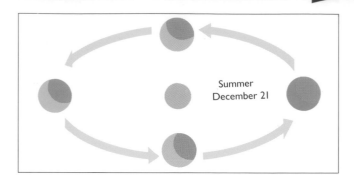

Summer
December 21

THE CONSTELLATIONS DESCRIBED

To find details of the major constellations shown on the map, refer to the pages indicated below.

Aries	64	Dorado	42	Perseus	67
Auriga	44	Eridanus	66	Phoenix	62
Cancer	46	Fornax	66	Pisces	63
Canis Major	44	Gemini	46	Piscis Austrinus	61
Canis Minor	46	Grus	60	Puppis	47
Carina	42	Hydra	48, 50	Sculptor	63
Centaurus	82	Lepus	45	Taurus	92
Cetus	64	Monoceros	47	Triangulum	65
Columba	45	Orion	86	Tucana	42
Crux	82				

USING THIS MAP

• This map represents the dome of the heavens at 11 p.m. local time on the date of the summer solstice, December 21.

• Because the stars rise about four minutes earlier each night, you will see a similar view of the heavens on December 14 at 11.30 p.m. and on December 28 at 10.30 p.m.

• The center of the map is your zenith. The edge of the map is your horizon. You may be able to see a little more or a little less north or south than the map shows, because your horizon depends on your latitude.

Looking north: To locate the constellations in the northern part of the sky, face north. The Sun will have set on your left. Hold the map in front of you with NORTH at the bottom. The lower half of the map now represents the part of the sky in front of you.

Looking south: To locate the constellations in the southern part of the sky, face south. The Sun will have set on your right. Turn the map upside-down so that SOUTH is at the bottom. The lower half of the map now represents the part of the sky in front of you.

Looking west

This hemisphere map shows the main constellations you would see looking west at about 11 p.m. local time on December 21. They may appear slightly higher or slightly lower in the sky, depending on your exact latitude.

Rather a bare sky. Aldebaran, the Hyades, and the Pleiades lie in the north.

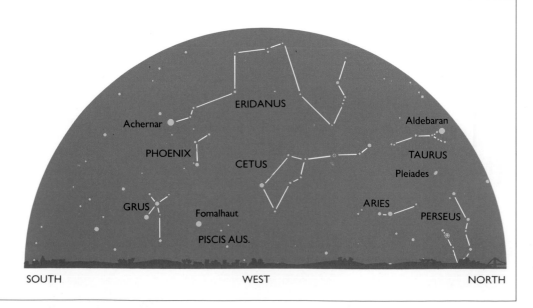

SOUTH   WEST   NORTH

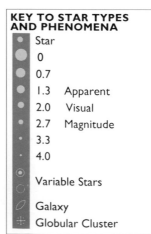

KEY TO STAR TYPES AND PHENOMENA

●	Star	
●	0	
●	0.7	
●	1.3	Apparent
●	2.0	Visual
●	2.7	Magnitude
●	3.3	
·	4.0	
◉	Variable Stars	
⬭	Galaxy	
✦	Globular Cluster	

December 21

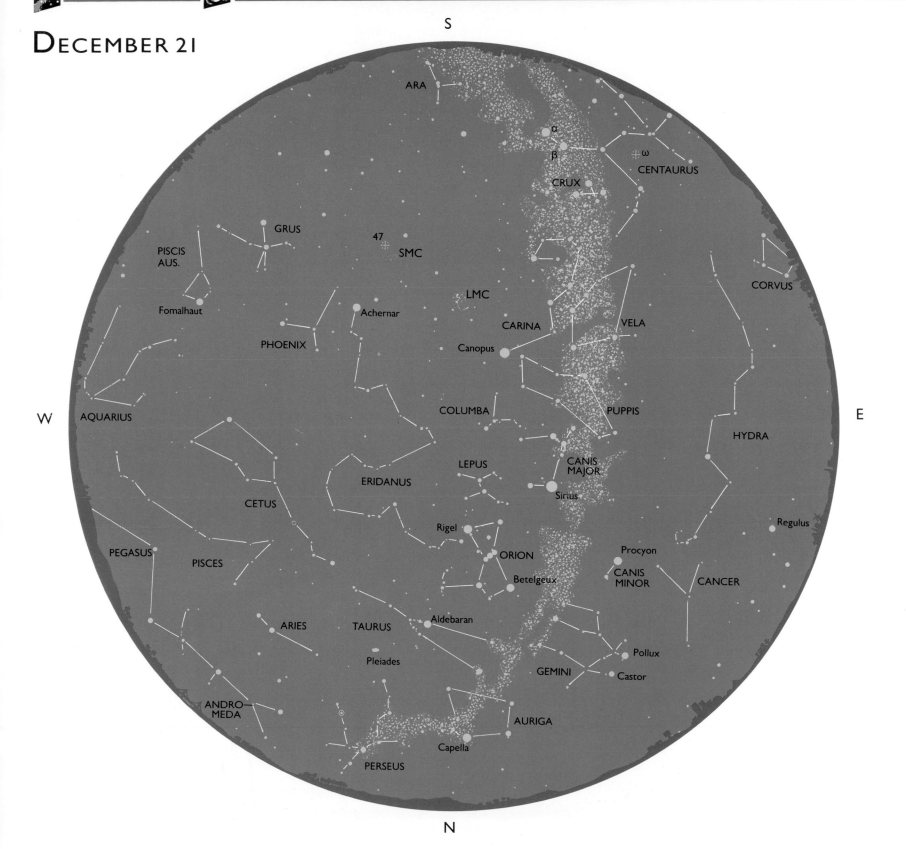

S

ARA

α

β

ω
CENTAURUS

CRUX

GRUS

PISCIS
AUS.

47
SMC

LMC

CORVUS

Fomalhaut

Achernar

CARINA

VELA

PHOENIX

Canopus

W

AQUARIUS

COLUMBA

PUPPIS

HYDRA

E

CETUS

ERIDANUS

LEPUS

CANIS
MAJOR

Sirius

Regulus

PEGASUS

Rigel

Procyon

PISCES

ORION

CANIS
MINOR

CANCER

Betelgeux

ARIES

TAURUS

Aldebaran

Pollux

Pleiades

GEMINI

Castor

ANDRO-
MEDA

AURIGA

Capella

PERSEUS

N

SOUTHERN HEMISPHERE FALL

The bright starscapes of winter no longer dominate the northern aspect of the sky in the late evening. Orion, Gemini, Canis Major, and Canis Minor are fast slipping away in the west.

It is Leo that now occupies center stage, with its hook-like pattern of stars "hanging" from the first-magnitude Regulus. Two other bright stars lie to the east. Spica (Virgo) at about the same elevation as Regulus (Leo) and Arcturus (Boötes) lower down. The remainder of the sky is relatively barren, because of the sprawling constellation Hydra near the zenith, which has no bright stars at all.

Turning to the southern aspect, this is occupied by a great swathe of the Milky Way, with more than its fair share of stellar delights. These range from the Southern Cross (Crux) and Alpha and Beta Centauri (Centaurus) high up, to the wickedly curved tail of Scorpius, the Scorpion, which is just appearing over the eastern horizon and will become one of winter's great pleasures.

The longer arm of the Southern Cross points almost to the celestial south pole. But there is no convenient bright star to mark the pole's position as there is in the Northern Hemisphere (Polaris, the Pole Star).

But the Cross does point more or less in the direction of the Small Magellanic Cloud (Tucana), located on the other side of the pole. The Large Magellanic Cloud (Dorado) is located higher up, in the direction of the brilliant Canopus (Carina). The two Clouds form a triangle with Achernar, the only bright star in Eridanus, recently risen above the horizon.

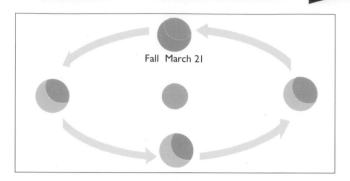

Fall March 21

THE CONSTELLATIONS DESCRIBED

To find details of the major constellations shown on the map, refer to the pages indicated below.

Ara 42	Crux 82	Lynx 49
Auriga 44	Dorado 42	Monoceros 47
Boötes 52	Eridanus 66	Orion 86
Cancer 46	Fornax 66	Phoenix 62
Canes Venatici 50	Gemini 46	Puppis 47
Canis Major 44	Hydra 48, 50	Scorpius 90
Canis Minor 46	Leo 48	Serpens Caput 55
Carina 42	Lepus 45	Tucana 42
Centaurus 82	Libra 53	Ursa Major 94
Columba 45	Lupus 53	Virgo 96
Coma Berenices 52		

USING THIS MAP

• This map represents the dome of the heavens at 11 p.m. local time on the date of the March equinox, March 21.

• Because the stars rise about four minutes earlier each night, you will see a similar view of the heavens on March 14 at 11.30 p.m. and on March 28 at 10.30 p.m.

• The center of the map is your zenith. The edge of the map is your horizon. You may be able to see a little more or a little less north or south than the map shows, because your horizon depends on your latitude.

Looking north: To locate the constellations in the northern part of the sky, face north. The Sun will have set on your left. Hold the map in front of you with NORTH at the bottom. The lower half of the map now represents the part of the sky in front of you.

Looking south: To locate the constellations in the southern part of the sky, face south. The Sun will have set on your right. Turn the map upside-down so that SOUTH is at the bottom. The lower half of the map now represents the part of the sky in front of you.

Looking west

This hemisphere map shows the main constellations you would see looking west at about 11 p.m. local time on March 21. They may appear slightly higher or slightly lower in the sky, depending on your exact latitude.

An abundance of bright stars graces western skies this month, led by Sirius and Canopus.

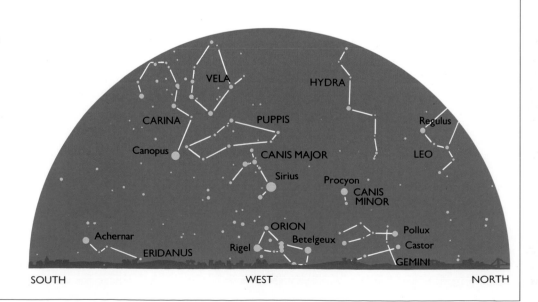

VELA · HYDRA · CARINA · PUPPIS · Regulus · Canopus · CANIS MAJOR · LEO · Sirius · Procyon · CANIS MINOR · ORION · Pollux · Achernar · Betelgeux · Castor · ERIDANUS · Rigel · GEMINI

SOUTH   WEST   NORTH

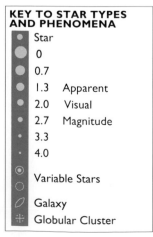

KEY TO STAR TYPES AND PHENOMENA

● Star	
● 0	
● 0.7	
● 1.3	Apparent
● 2.0	Visual
● 2.7	Magnitude
· 3.3	
· 4.0	
◉ Variable Stars	
◯ Galaxy	
⊘ Galaxy	
✦ Globular Cluster	

MARCH 21

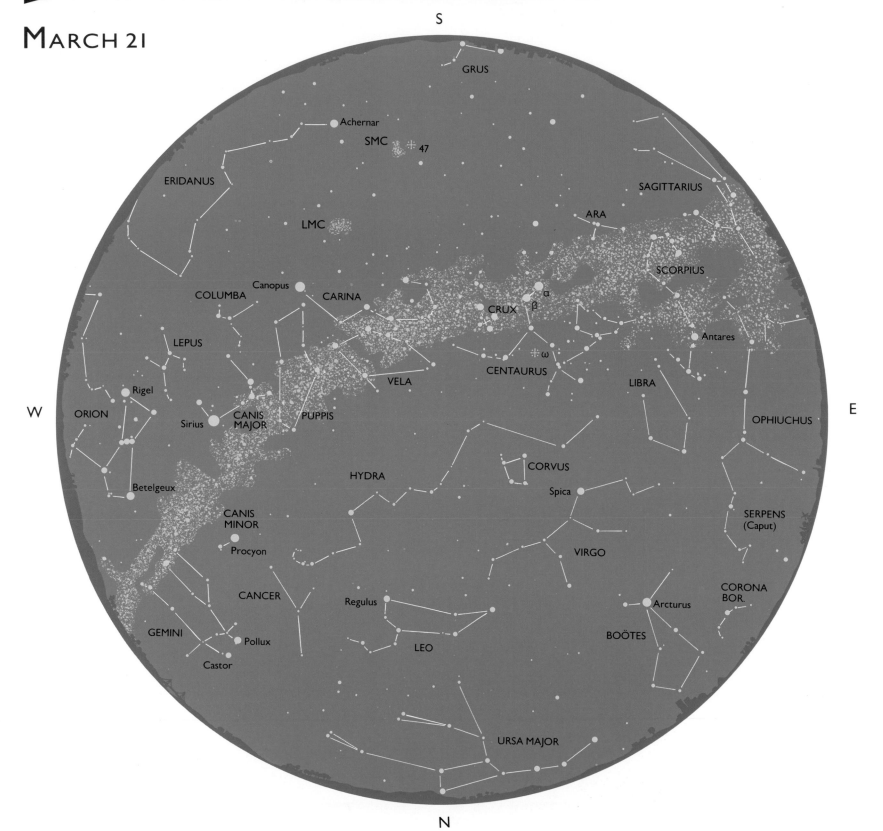

S

GRUS

Achernar

SMC ✽ 47

ERIDANUS

SAGITTARIUS

LMC

ARA

SCORPIUS

COLUMBA   Canopus   CARINA

CRUX   α
β

LEPUS   CENTAURUS   Antares

✽ ω

Rigel   VELA

ORION   PUPPIS   LIBRA

W   Sirius   CANIS
MAJOR   OPHIUCHUS   E

Betelgeux   CORVUS

HYDRA   Spica   SERPENS
(Caput)

CANIS
MINOR   VIRGO

Procyon   CORONA
BOR.

CANCER   Regulus   Arcturus

GEMINI   Pollux   LEO   BOÖTES

Castor

URSA MAJOR

N

SOUTHERN HEMISPHERE WINTER

The winter skies in the Southern Hemisphere are the stamping ground of the Scorpion (Scorpius), whose curved tail is poised ready to deal a deadly sting. The constellation is high overhead, instantly recognizable and breathtaking in its beauty. The brightest star, the red Antares, lies near the zenith.

Cheek-by-jowl with the Scorpion in what is the richest part of the Milky Way is Sagittarius. This constellation overflows with astronomical delights, which binoculars show to perfection. There are dense and brilliant star clouds, swirling nebulae, and scattered star clusters. The reason why the Milky Way appears so dense here is because the center of our Galaxy lies in this direction.

The northern aspect of the sky shows, south of Sagittarius and also in the Milky Way, the bright Altair (Aquila). Vega (Lyra) farther south and east rises later, while Deneb (Cygnus) rises later still. These three stars form a prominent winter triangle, which northern astronomers call the Summer Triangle to match their own season.

The higher of the only two prominent stars in the northwest is Spica (Virgo); the lower is Arcturus (Boötes). Close by is the half-circle of stars known as the Northern Crown (Corona Borealis), which is so much easier to place than its southern equivalent.

The southern aspect of the winter sky reveals the

Southern Cross (Crux) and its bright pointers Alpha and Beta Centauri (Centaurus) descending. Alpha Centauri is the nearest bright star to us, a "mere" 4.3 light-years, or some 25 million million miles (40 million million kilometers) away.

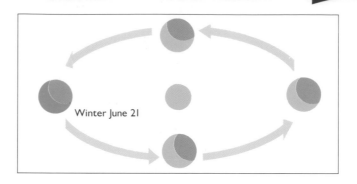

Winter June 21

THE CONSTELLATIONS DESCRIBED

To find details of the major constellations shown on the map, refer to the pages indicated below.

Aquarius 60	Delphinus 58	Phoenix 62
Aquila 58	Dorado 42	Piscis Austrinus 61
Ara 42	Eridanus 66	Puppis 47
Boötes 52	Grus 60	Sagittarius 88
Canes Venatici 50	Hercules 54	Scorpius 90
Capricornus 58	Hydra 48, 50	Scutum 57
Carina 42	Leo 48	Serpens Caput 55
Centaurus 82	Libra 53	Serpens Cauda 57
Coma Berenices 52	Lupus 53	Tucana 42
Corona Borealis 54	Lyra 56	Vela 49
Crux 82	Ophiuchus 56	Virgo 96

USING THIS MAP
● This map represents the dome of the heavens at 11 p.m. local time on the date of the winter solstice, June 21.
● Because the stars rise about four minutes earlier each night, you will see a similar view of the heavens on June 14 at 11.30 p.m. and on June 28 at 10.30 p.m.
● The center of the map is your zenith. The edge of the map is your horizon. You may be able to see a little more or a little less north or south than the map shows, because your horizon depends on your latitude.

Looking north: To locate the constellations in the northern part of the sky, face north. The Sun will have set on your left. Hold the map in front of you with NORTH at the bottom. The lower half of the map now represents the part of the sky in front of you.

Looking south: To locate the constellations in the southern part of the sky, face south. The Sun will have set on your right. Turn the map upside-down so that SOUTH is at the bottom. The lower half of the map now represents the part of the sky in front of you.

Looking west

This hemisphere map shows the main constellations you would see looking west at about 11 p.m. local time on June 21. They may appear slightly higher or slightly lower in the sky, depending on your exact latitude.

Alpha, Beta, and Omega Centauri and Crux are bright in the south; Arcturus in the north.

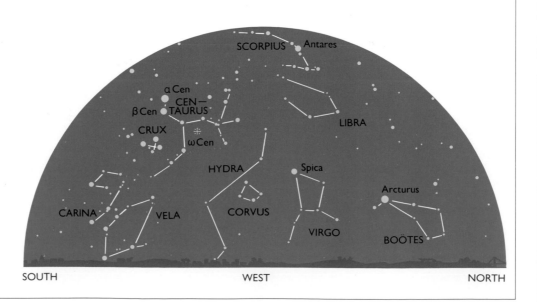

SOUTH   WEST   NORTH

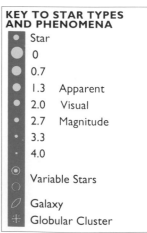

KEY TO STAR TYPES AND PHENOMENA

● Star	
● 0	
● 0.7	
● 1.3	Apparent
● 2.0	Visual
● 2.7	Magnitude
● 3.3	
· 4.0	
⊙ Variable Stars	
○	
⬭ Galaxy	
✳ Globular Cluster	

June 21

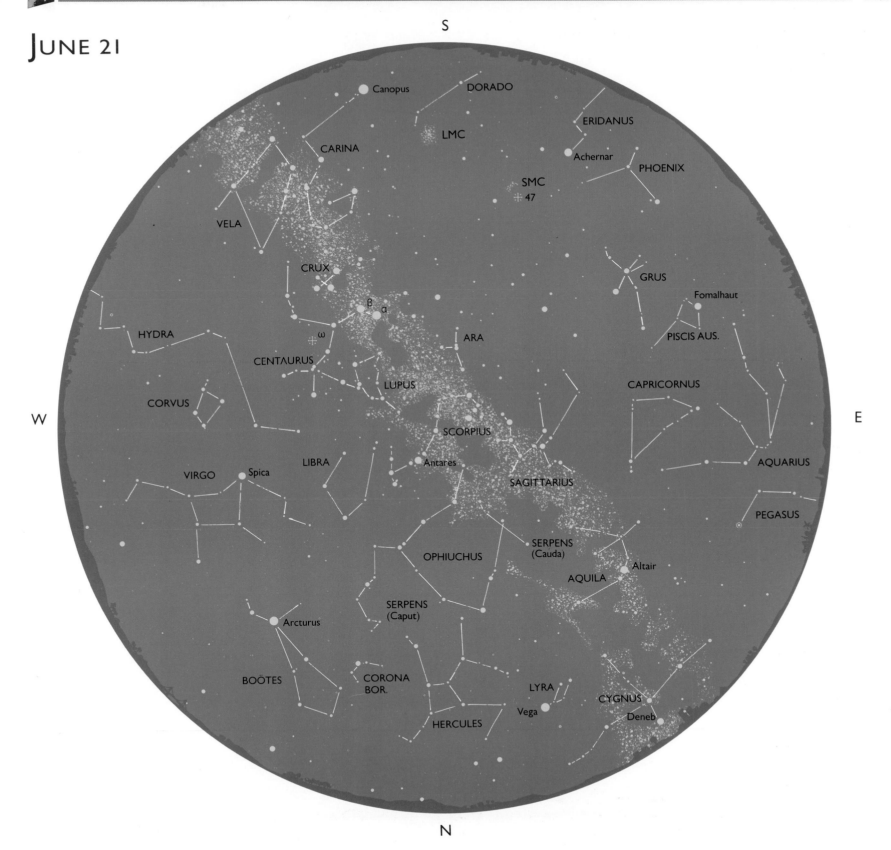

S

W

E

N

Canopus   DORADO
CARINA   LMC   ERIDANUS
Achernar   PHOENIX
VELA   SMC
‡ 47
CRUX   GRUS
β   α   Fomalhaut
HYDRA   ω   ARA   PISCIS AUS.
CENTAURUS   LUPUS
CAPRICORNUS
CORVUS   SCORPIUS   AQUARIUS
LIBRA   Antares
VIRGO   Spica   SAGITTARIUS   PEGASUS
OPHIUCHUS   SERPENS   AQUILA   Altair
(Cauda)
SERPENS
(Caput)
BOÖTES   Arcturus   CORONA   LYRA   CYGNUS
BOR.   Vega   Deneb
HERCULES

SOUTHERN HEMISPHERE SPRING

The great delights of winter skies, Sagittarius and Scorpius, are now seen in the southwest. They are still well placed for viewing, but in the next few weeks they will disappear from the late evening skies.

Still looking south, the Southern Cross (Crux) and its pointers, Alpha and Beta Centauri (Centaurus), are now low down on the horizon. So is the "False Cross," formed by a pair of stars in Vela and a pair in Carina.

The Magellanic Clouds are particularly well placed for observation, the Small Cloud appearing nearly due south and the Large Cloud (Dorado) more to the east and lower down.

The two Clouds are more or less in line with the brilliant Canopus (Carina), a star only exceeded in brightness by Sirius (Canis Major), which is just rising (or about to rise, depending on the observer's latitude) above the eastern horizon.

The northern aspect of the spring sky shows the Square of Pegasus prominent. The showpiece constellations of summer, Taurus and Orion, are about to rise in the east. The Pleiades, or Seven Sisters, star cluster may well be already above the horizon.

High overhead, near the zenith, Fomalhaut (Piscis Austrinus) is conspicuous, not because it is outstandingly bright but because it appears in a relatively barren part of the heavens.

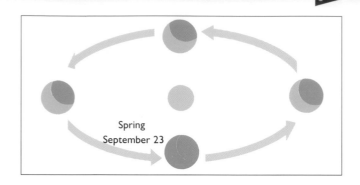

Spring
September 23

THE CONSTELLATIONS DESCRIBED

To find details of the major constellations shown on the map, refer to the pages indicated below.

Andromeda 78	Columba 45	Ophiuchus 56
Aquarius 60	Crux 82	Pegasus 62
Aquila 58	Cygnus 84	Phoenix 62
Ara 42	Delphinus 58	Pisces 63
Aries 64	Dorado 42	Sagittarius 88
Capricornus 58	Eridanus 66	Serpens Cauda 57
Carina 42	Grus 60	Scorpius 90
Centaurus 82	Lupus 53	Scutum 57
Cetus 64	Lyra 56	

In the northwest, two or even three stars of the "winter triangle" may still be visible. Altair (Aquila) is relatively high in the sky, while Deneb (Cygnus) is close to setting, and Vega (Lyra) may have set already.

USING THIS MAP

● This map represents the dome of the heavens at 11 p.m. local time on the date of the September equinox, September 23.

● Because the stars rise about four minutes earlier each night, you will see a similar view of the heavens on September 16 at 11.30 p.m. and on September 30 at 10.30 p.m.

● The center of the map is your zenith. The edge of the map is your horizon. You may be able to see a little more or a little less north or south than the map shows, because your horizon depends on your latitude.

Looking north: To locate the constellations in the northern part of the sky, face north. The Sun will have set on your left. Hold the map in front of you with NORTH at the bottom. The lower half of the map now represents the part of the sky in front of you.

Looking south: To locate the constellations in the southern part of the sky, face south. The Sun will have set on your right. Turn the map upside-down so that SOUTH is at the bottom. The lower half of the map now represents the part of the sky in front of you.

Looking west

This hemisphere map shows the main constellations you would see looking west at about 11 p.m. local time on September 23. They may appear slightly higher or slightly lower in the sky, depending on your exact latitude.

The Milky Way is almost horizontal and low down, and boasts delights galore.

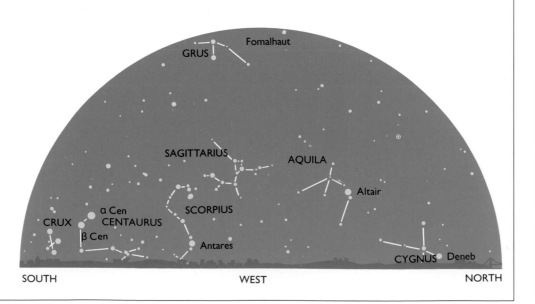

Fomalhaut

GRUS

SAGITTARIUS

AQUILA

Altair

α Cen
CENTAURUS
SCORPIUS

CRUX

β Cen

Antares

CYGNUS   Deneb

SOUTH   WEST   NORTH

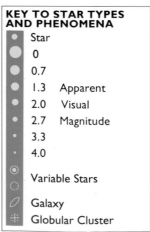

KEY TO STAR TYPES AND PHENOMENA

●	Star
●	0
●	0.7
●	1.3 Apparent
●	2.0 Visual
●	2.7 Magnitude
·	3.3
·	4.0
⊙	Variable Stars
○	
🍃	Galaxy
⊕	Globular Cluster

September 23

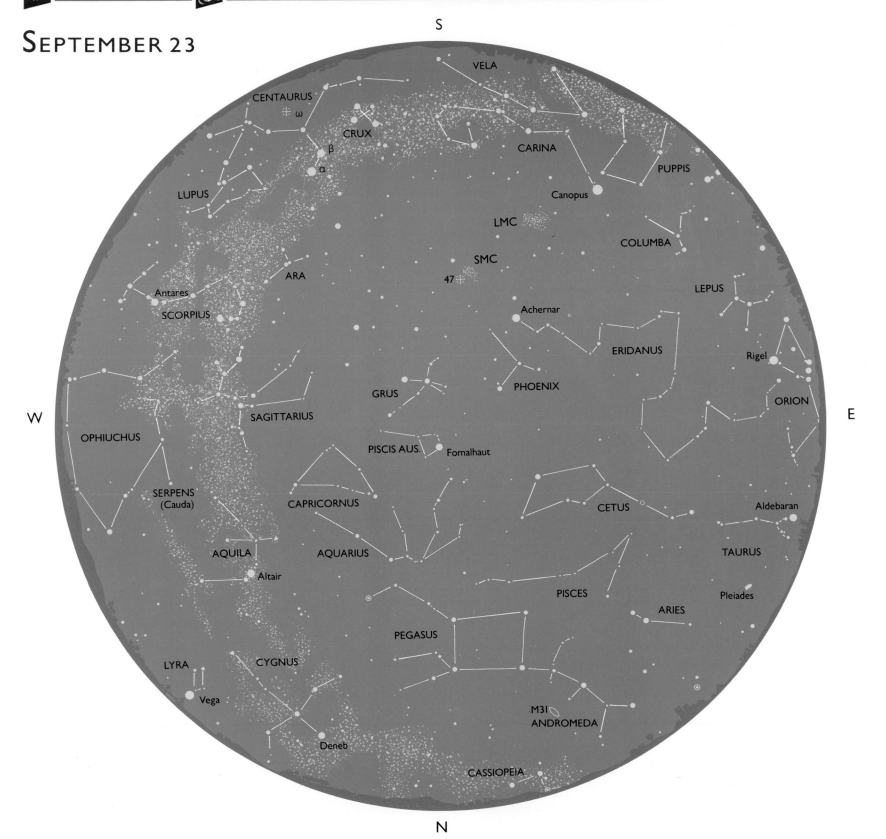

S

CENTAURUS
✳ ω
VELA
CRUX
β
α
CARINA
PUPPIS
LUPUS
Canopus
LMC
COLUMBA
ARA
SMC
LEPUS
47 ✳
Antares
Achernar
SCORPIUS
ERIDANUS
Rigel
PHOENIX
GRUS
ORION
W
E
SAGITTARIUS
OPHIUCHUS
PISCIS AUS.
Fomalhaut
SERPENS
(Cauda)
CETUS
Aldebaran
CAPRICORNUS
AQUILA
AQUARIUS
TAURUS
Altair
PISCES
Pleiades
ARIES
PEGASUS
LYRA
CYGNUS
Vega
M31
ANDROMEDA
Deneb
CASSIOPEIA

N

THE CONSTELLATIONS

Observational astronomy hinges on the star patterns we call the constellations, which enable us quickly to find a particular star in the heavens. The constellations do not appear to change century after century. Their stars seem to be fixed in position and travel as a group through space. This, however, is an illusion.

▼ Constellations of the Northern Hemisphere, featured in an ancient astronomy book. The artist has used his imagination to create some interesting characters. Behind each one, there is a fascinating mythological story of love, lust, jealousy, hate, heroism, and sacrifice.

The stars in the constellations are not in fact grouped together in space. They are quite independent of one another and are usually separated by vast distances. We see them together in the sky merely because they happen to lie in the same direction in space. For example, of the seven brightest stars in the familiar constellation Orion, the closest (Gamma) lies about 300 light-years away, while the most distant (Kappa) lies over 1,800 light-years away.

Again, the stars in the constellations appear to be fixed in position, but they are not. They are all moving through space, at different speeds and in different directions. But they lie so very far away from us that, except in a few cases, their movement is imperceptible. Only after tens of thousands of years will the star patterns we see today show noticeable changes.

Naming the constellations

Altogether, there are 88 recognized constellations (see opposite). Many of them date back at least 5,000 years. We know this from the records kept by ancient stargazers in the Middle East – in Babylon and Egypt.

But it is to the ancient Greeks that we are indebted for the names of most of the constellations that we use today, although we use the Latin form of the Greek names. We also often call the constellations by their translated names. For example, the constellation Ursa Major translates into English as the Great Bear, Leo as the Lion, and Cygnus as the Swan.

Myths and monsters

The Greeks named the constellations after gods, heroes, creatures, and objects that featured in their mythology, matching the star patterns to the mythological figures. For many constellations you need an active imagination to see the resemblance between the figure and the star pattern it is meant to represent.

However, in some instances, the constellation patterns really do look like the figures they are intended to represent. Scorpius, for example, really does look like a scorpion, with a wicked curved tail. Cygnus really does look like a flying swan, with its long neck outstretched. Ursa Major can with a little imagination turn into a bear, and Leo into a lion.

Ptolemy and Bayer

The Greeks recognized some 48 constellations. The last great ancient Greek astronomer Ptolemy listed them in his celebrated encyclopedia of astronomy and mathematics (*circa* A.D.140), which has come down to us in its Arab translation, known as the *Almagest*.

Forty constellations were later added, notably by the German astronomers Johann Bayer (in 1603) and Johann Hevelius (in 1690) and the French astronomer Nicolas Lacaille (in 1752).

Bayer also introduced the system of using letters of the Greek alphabet to identify the stars in a constellation. He designated the brightest star Alpha (α), the second brightest Beta (β), the third brightest Gamma (γ), and so on. Bayer was not always accurate in grading the stars in this way; nevertheless, astronomers still follow the designations he adopted.

1 Andromeda
2 Antlia, Air Pump
3 Apus, Bird of Paradise
4 Aquarius, Water-Bearer
5 Aquila, Eagle
6 Ara, Altar
7 Aries, Ram
8 Auriga, Charioteer
9 Boötes, Herdsman
10 Caelum, Graving Tool
11 Camelopardalis, Giraffe
12 Cancer, Crab
13 Canes Venatici, Hunting Dogs
14 Canis Major, Great Dog
15 Canis Minor, Little Dog
16 Capricornus, Sea Goat
17 Carina, Keel
18 Cassiopeia
19 Centaurus, Centaur
20 Cepheus
21 Cetus, Whale
22 Chamaeleon, Chameleon
23 Circinus, Compasses
24 Columba, Dove
25 Coma Berenices, Berenice's
   Hair

26 Corona Australis, Southern
   Crown
27 Corona Borealis, Northern
   Crown
28 Corvus, Crow
29 Crater, Cup
30 Crux, Southern Cross
31 Cygnus, Swan
32 Delphinus, Dolphin
33 Dorado, Swordfish
34 Draco, Dragon
35 Equuleus, Foal
36 Eridanus
37 Fornax, Furnace
38 Gemini, Twins
39 Grus, Crane
40 Hercules
41 Horologium, Clock
42 Hydra, Water Snake
43 Hydrus, Little Snake
44 Indus, Indian
45 Lacerta, Lizard
46 Leo, Lion
47 Leo Minor, Little Lion
48 Lepus, Hare
49 Libra, Scales

Northern Hemisphere

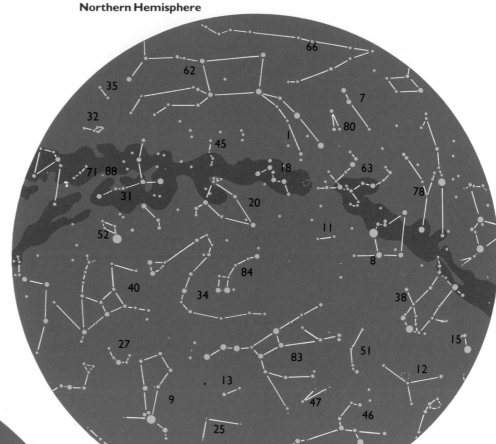

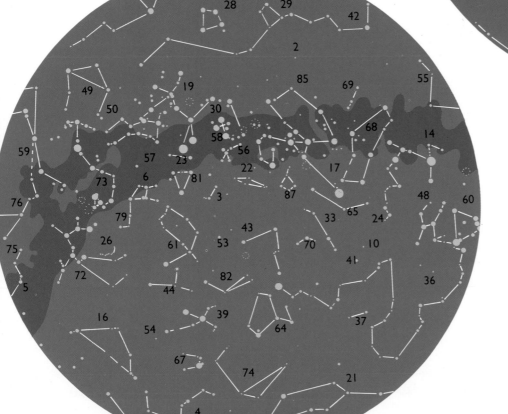

Southern Hemisphere

50 Lupus, Wolf
51 Lynx, Lynx
52 Lyra, Lyre
53 Mensa, Table
54 Microscopium, Microscope
55 Monoceros, Unicorn
56 Musca, Fly
57 Norma, Rule
58 Octans, Octant
59 Ophiuchus, Serpent-Bearer
60 Orion
61 Pavo, Peacock
62 Pegasus, Flying Horse
63 Perseus
64 Phoenix, Phoenix
65 Pictor, Painter
66 Pisces, Fishes
67 Piscis Austrinus, Southern Fish
68 Puppis, Poop
69 Pyxis, Compass

70 Reticulum, Net
71 Sagitta, Arrow
72 Sagittarius, Archer
73 Scorpius, Scorpion
74 Sculptor, Sculptor
75 Scutum, Shield
76 Serpens, Serpent
77 Sextans, Sextant
78 Taurus, Bull
79 Telescopium, Telescope
80 Triangulum, Triangle
81 Triangulum Australe, Southern
   Triangle
82 Tucana, Toucan
83 Ursa Major, Great Bear
84 Ursa Minor, Little Bear
85 Vela, Sails
86 Virgo, Virgin
87 Volans, Flying Fish
88 Vulpecula, Fox

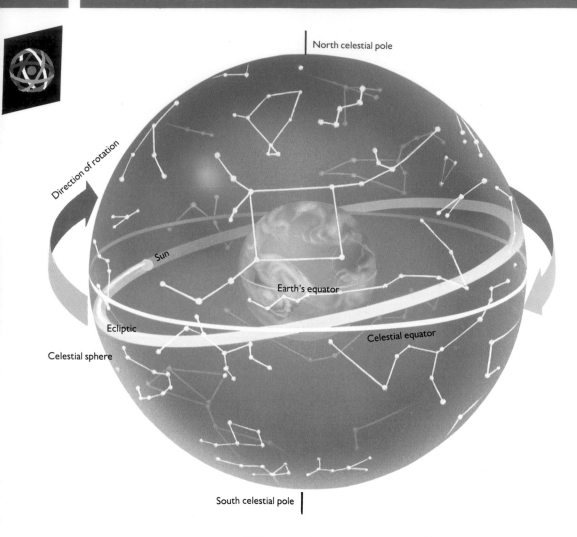

North celestial pole

Direction of rotation

Sun

Earth's equator

Ecliptic

Celestial sphere

Celestial equator

South celestial pole

▲ The concept of the celestial sphere: the stars are stuck on the inside of the sphere in fixed positions. The Earth is at the center and is stationary; the celestial sphere rotates around it from east to west.

The *celestial equator* is a circle around the celestial sphere halfway between the celestial poles; it is a projection of the Earth's Equator. The stars follow paths parallel with the celestial equator.

An observer's view of the celestial sphere is bounded all around by the *horizon*, where the horizontal plane through the observer's position meets the sphere. The point on the celestial sphere directly above the observer's head is the *zenith*. The equivalent point on the sphere directly beneath the observer is the *nadir*.

The great circle passing through the north celestial pole, the zenith, and the south celestial pole is the *meridian*. Stars *culminate*, or reach their highest altitude, on the meridian.

The *ecliptic* is the apparent path of the Sun around the celestial sphere during the year. The Sun crosses the celestial equator on March 21, moving north; and on September 23, moving south. On these dates, day and night are exactly 12 hours long all over the world. That is why they are called the equinoxes ("equal nights").

▼ Observers in middle latitudes will see the north celestial pole halfway between the zenith and the horizon. They will see the stars rise in the east, arc through the sky, and then set in the west. The stars near the north celestial pole will describe complete circles and will always be visible.

► Star trails over the Roque de los Muchachos Observatory on La Palma in the Canary Islands. The two domes house the 39-inch (1-meter) Jacobus Kapteyn Telescope *(left)* and the 98-inch (2.5-meter) Isaac Newton Telescope. The view shows stars circling the north celestial pole.

THE CELESTIAL SPHERE

We noted earlier that here on Earth we seem to be at the center of a great enveloping heavenly or celestial sphere, which rotates around us from east to west carrying the stars and the other heavenly bodies with it.

Although we know that there is no such thing as a celestial sphere and that the motion of the heavenly bodies is an apparent one, created by the Earth spinning on its axist from west to east, the idea is still useful to observational astronomy. The two diagrams on this page show features of the sphere that astronomers frequently refer to.

The *north and south celestial poles* are points on the celestial sphere that lie directly above the Earth's North and South Poles respectively. The sphere appears to rotate on an axis through the celestial poles. The star Polaris lies very close to the north celestial pole and is called the Pole Star. There is no convenient bright star near the south celestial pole.

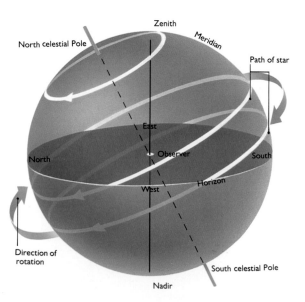

Zenith

Meridian

North celestial Pole

Path of star

East

Observer

South

North

West

Horizon

Direction of rotation

South celestial Pole

Nadir

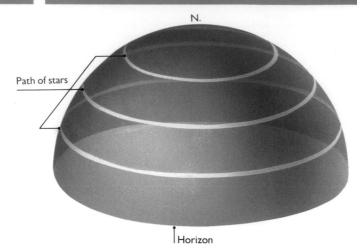

N.

Path of stars

Horizon

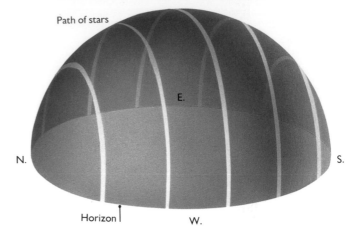

Path of stars

E.

N.

S.

Horizon

W.

Observers at the North and South Poles and on the Equator view the heavens in quite a different way from everyone else. At the North Pole, for example (*left*), they will see the stars moving parallel with the horizon. On the Equator (*right*), the stars rise and set vertically. So if they look east or west, they will see the stars move straight up or straight down.

Celestial time

The spinning of the Earth on its axis provides the basis for measuring time. Our day of 24 hours is the time it takes the Earth to spin around once and return to the same position in the sky *relative to the Sun*. We call this, our ordinary time, "solar time."

At first sight, we might also expect the celestial sphere to rotate around us in 24 hours. But it does not. If you note the time when a certain star rises above the horizon on successive days, you will find that on the second day the star rises four minutes earlier than on the first, and that on the third day it rises four minutes earlier still, and so on.

In other words, the celestial sphere rotates once around the Earth in 23 hours 56 minutes. Or, put the other way, the Earth rotates once *relative to the stars* in 23 hours 56 minutes. This is the Earth's true period of rotation and

forms the basis of what we call "sidereal time," or time *relative to the stars*.

When they are observing, astronomers use sidereal time. Then the stars rise, culminate, and set at the same time. In other words, they are in the same position in the sky at the same time.

Pinpointing the stars

Many of the brightest stars are easily located because they form part of the recognizable pattern of a constellation. But other stars are more difficult to pinpoint accurately. Astronomers therefore follow geographers on Earth and locate stars by a stellar "map reference." The method they use is exactly analogous to the system of terrestrial latitude and longitude.

The latitude of a point on Earth is the angular distance of the point north or south of the Equator. It is

▼ The Sun crosses the celestial equator at two points on its annual path (ecliptic) around the celestial sphere. It crosses at the vernal equinox (March 21) at a point called the First Point in Aries, and at the autumnal equinox (September 23) at the First Point in Libra.

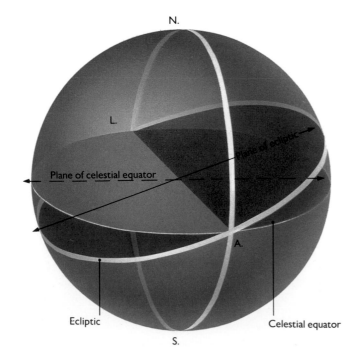

N.

L.

Plane of ecliptic

Plane of celestial equator

A.

Ecliptic

Celestial equator

S.

Celestial coordinates

We pinpoint the position of a star on the celestial sphere by a system of celestial latitude (declination) and longitude (right ascension). The declination (δ) is the angular distance of the star from the celestial equator. It is measured in degrees.

The right ascension (RA) is the angular distance of the star along the celestial equator from the First Point in Aries. This is the point where the Sun's path (the ecliptic) crosses the celestial equator in spring. Right ascension is measured in hours and minutes.

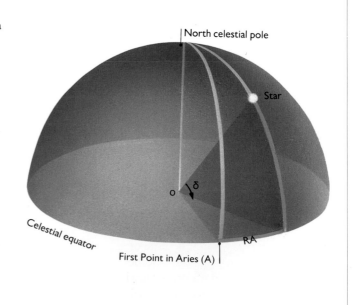

North celestial pole

Star

o δ

Celestial equator

First Point in Aries (A)

RA

measured in degrees and minutes. The celestial latitude, or declination, of a star is the angular distance the star is north or south of the celestial equator. It, too, is measured in degrees and minutes. Declinations north of the celestial equator go from 0° at the equator to +90° at the north celestial pole. Declinations south of the equator go from 0° to −90° at the south celestial pole.

Celestial longitude

The longitude of a point on Earth is the angular distance measured east along the Equator from the meridian passing through Greenwich, in England, to the meridian passing through the point. Like latitude, it is measured in degrees and minutes.

Celestial longitude, or right ascension (RA), is also an angular distance measured along the celestial equator. The starting point for measuring right ascension is the First Point in Aries, the point where the ecliptic crosses the celestial equator on March 21, the spring equinox.

So the right ascension of a star is the angular distance measured east from the First Point in Aries to the great circle passing through the star. It is not measured in degrees, however. It is measured in units of sidereal time. The reasoning behind this is that the celestial sphere rotates once every 24 sidereal hours; therefore 24 hours is equivalent to one complete circle of 360 degrees. One hour is equivalent to 15 degrees. Or in other words, the celestial sphere turns around 15 degrees in one hour.

This all sounds rather complicated, but this is more because of the terminology rather than the concept. If you can map-read on Earth using latitude and longitude, you will soon get the hang of map-reading on the celestial sphere using declination and right ascension.

Planispheres

A useful device to carry with you when you are stargazing is a planisphere. It provides a quick and convenient way of finding and identifying the stars visible on any date and at any time of the night. Since the stars you see depend on your location on Earth, different planispheres are available for different latitudes.

A planisphere consists of two parts, a base and a movable disk that can rotate over the base. The base carries a map of all the stars visible from the given latitude on Earth during the year. A scale of months and days is marked around the edge. The movable disk has a window that reveals the stars visible above the horizon at a particular time. A scale showing the time of day (local time) in hours is marked around the edge of the movable disk.

Matching the hour of observation with the date brings the stars visible at that time into the window.

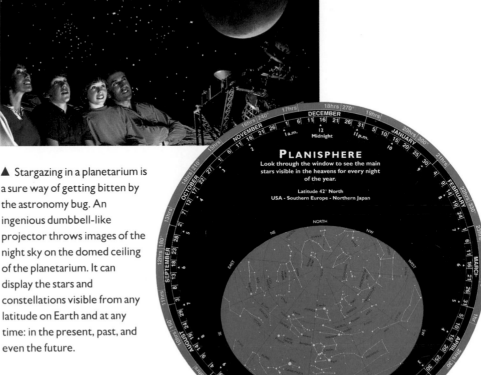

▲ Stargazing in a planetarium is a sure way of getting bitten by the astronomy bug. An ingenious dumbbell-like projector throws images of the night sky on the domed ceiling of the planetarium. It can display the stars and constellations visible from any latitude on Earth and at any time: in the present, past, and even the future.

▶ Planispheres for Earth latitudes 42° North (top) and 35° South. The former is suitable for most of the U.S. and southern Europe; the latter for Australia and New Zealand. Note that the scales on these two planispheres go in opposite directions. This is because the northern planisphere is centered on the northern celestial pole, and the southern one is centered on the southern celestial pole. So looking toward the poles, the sky appears to rotate in opposite directions.

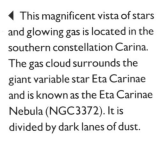

◀ This magnificent vista of stars and glowing gas is located in the southern constellation Carina. The gas cloud surrounds the giant variable star Eta Carinae and is known as the Eta Carinae Nebula (NGC3372). It is divided by dark lanes of dust.

2

MAPPING THE STARS

As you saw in the first chapter, different constellations come into view season by season as the Earth makes its annual journey around the Sun. This chapter takes a closer look at the constellations as, month by month, they take turns to occupy center stage in the late evening skies.

Views of the night sky over the year are presented as a series of monthly maps, from January to December. These maps represent segments of the celestial sphere. Each map shows all the constellations visible in that segment at that particular time. But which constellations an observer will see depends on the observer's latitude, or precise location on Earth. An observer in Australia, in the Southern Hemisphere, will not be able to see some of the far northern constellations that an observer in Canada will see, and vice versa.

Two additional star maps are centered on the celestial poles, representing the "top" and "bottom" of the celestial sphere. To far northern and far southern observers, part of these regions of the sky will always be visible throughout the year.

The Monthly Maps

Fourteen star maps appear in this chapter – one for each month of the year, plus one each for the north and south polar regions. Together, the maps cover the whole of the celestial sphere, with plenty of overlap. As with all maps, there is a certain amount of distortion, for they are representing a curved surface on flat paper.

▼ The European astronomy satellite Hipparchos (high-precision parallax collecting satellite), which has established the positions of more than 120,000 stars with unprecedented accuracy. It has brought a new dimension to astrometry, or positional astronomy.

The maps are drawn with a grid of celestial latitude and longitude, or "declination" and "right ascension." Declination is measured in degrees north (+) or south (−) of the celestial equator; right ascension is measured in hours and minutes of sidereal time. This grid provides a reference for locating stars on the celestial sphere. It can be used to locate other stars that are not included on these maps.

These maps feature stars down to the fifth magnitude, slightly above naked-eye visibility. Stars fainter than this are shown if they are of particular interest.

The stars are labeled according to the Bayer system with letters of the Greek alphabet, in which α (alpha) is the brightest in the constellation, β (beta) the next brightest, and so on. On these maps, usually only the stars featured in the text are labeled, making them easier to locate. In the text the letters are spelled out, as in Alpha Centauri. As an *aide memoire*, the complete Greek alphabet is given in the table below.

Many of the brightest stars also have a proper name, such as Sirius for Alpha Canis Majoris. Proper names are given for the best known of these.

Some variable stars are identified by their Bayer letter, but others are identified according to a different system. The first variable discovered in a constellation is denoted R, the next S, and so on. After Z, the variables become RR, RS, and so on. On the maps, two symbols are used for variables: ⊙ for variables that remain visible to the naked eye at minimum, and ○ for variables that fall below naked-eye visibility.

Deep-sky objects

By this we mean nebulae, open clusters, globular clusters, and galaxies. These are marked on the maps with their own symbols, and they are usually identified either by an M number or just by a number.

The M number is the Messier number, or the number assigned to the particular object in a catalog of more than 100 nebulous objects drawn up by the French astronomer Charles Messier in the 18th century. M42 in Orion is the Orion Nebula. Messier was an ardent comet-hunter and drew up his catalog pinpointing nebulae and clusters that might be confused with new comets.

A number by itself next to a deep-sky object symbol denotes the NGC number of the object. This is the number in the *New General Catalog of Nebulae and Clusters of Stars* compiled by the Danish astronomer Johann Dreyer and first published in 1888. The Messier numbers also have their equivalent NGC numbers: M42, for example, is NGC1976. But on the maps we show only the Messier number.

► Stars pack together densely in the Milky Way, as here in Monoceros. The rose-petal form of the cloud of glowing gas we see here earns it the name of the Rosette Nebula. The open star cluster at its center is NGC2244, easily seen in binoculars.

This diagram shows how the monthly star maps that follow relate to the celestial sphere. The celestial sphere is shown "exploded" into 12 segments – one for each month of the year.

Each segment covers two hours of Right Ascension. It forms the central part of one of the monthly maps (see right). The stars, of course, appear on the inner surface of the segment.

The north polar and south polar star maps are derived from the northern and southern "caps" of the celestial sphere, centered on the celestial poles.

North celestial pole

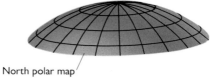
North polar map

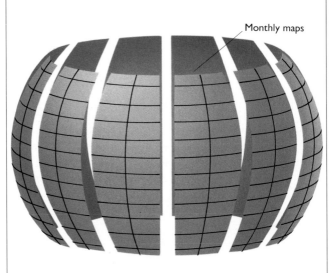
Monthly maps

South polar map

South celestial pole

THE GREEK ALPHABET

α Alpha	ι Iota	ρ Rho
β Beta	κ Kappa	σ Sigma
γ Gamma	λ Lambda	τ Tau
δ Delta	μ Mu	υ Upsilon
ε Epsilon	ν Nu	φ Phi
ζ Zeta	ξ Xi	χ Chi
η Eta	ο Omicron	ψ Psi
θ Theta	π Pi	ω Omega

▼ This is a smaller version of one of the monthly maps (October), with annotations to point out salient features.

A This lighter area shows the position of the Milky Way.

B Symbol for a planetary nebula. The number is its NGC number.

C Symbol for an open cluster. The number is its NGC number.

D Symbol for a galaxy. The M number is its number in the Messier catalog.

E This dashed line shows the ecliptic, the apparent path of the Sun across the celestial sphere.

F The line of zero declination, marking the celestial equator.

G The Greek letter Beta is the Bayer letter, which indicates that this star is the second-brightest in the constellation.

H Star name. Some of the best-known stars are labeled with their name as well as their Bayer letter.

▶ Also included on each spread with the monthly star map are two "sky views" like this. One shows what an observer in mid-latitudes in the Northern Hemisphere would see looking south at about 11 p.m. local time in the first week of the month. The other shows what an observer in the Southern Hemisphere would see looking north at the same time.

I Name of the constellation in Latin.

J Scale of declination, or celestial latitude. The minus sign applies in the southern hemisphere.

K Scale of right ascension, or celestial longitude.

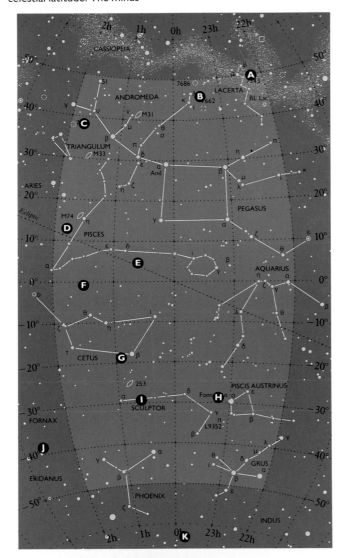

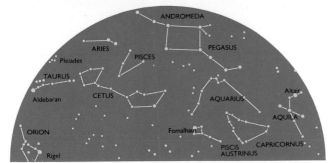

North Polar Stars

The north polar region is centered on Polaris, the Pole Star, which moves scarcely at all in the sky. Ursa Major is the outstanding constellation and is circumpolar – always visible – to observers north of latitude 40°.

▲ Stars trail around the north celestial pole in this long-exposure photograph taken with an ordinary tripod-mounted camera. The short arc in the center was made by Polaris, which is less than one degree away from the pole.

Andromeda *See page 78.*

Auriga, the Charioteer *See page 44.*

Boötes, the Herdsman *See page 52.*

Cassiopeia *See page 80.*

Cepheus
This is quite a faint constellation, probably best found by using Alpha and Beta Cassiopeiae as pointers. By far the most spectacular star in the constellation is Mu, which William Herschel dubbed the Garnet Star. It is noticeably red even to the naked eye and looks really beautiful in binoculars. Mu is a variable star of the Mira type, varying between about magnitudes 3.5 and 5 over about two years.

Another interesting star is Delta, also a variable. It varies with absolute precision between magnitudes 3.5 and 4.4 in 5.4 days. It is the prototype star of the regular-as-clockwork variables known as the Cepheids. Delta is also a double star, with a bluish companion.

Cygnus, the Swan *See page 84.*

Draco, the Dragon
This sprawling constellation winds nearly halfway around the north celestial pole. Interestingly, Alpha, also called Thuban, used to be the Pole Star about 4,500 years ago. A passage in the Great Pyramid of Giza, constructed at about that time, was aligned on the star. The location of the celestial poles changes slowly because of precession, a kind of wobbling of the Earth's axis.

Among the several fine double stars in Draco is Nu, a pair of white stars of identical brightness. Eta, Psi, and Xi are also double. Enclosed by the "neck" of the Dragon is one of the finest planetary nebulae in the heavens, 37 Draconis (NGC6543). It presents a noticeably blue disk, although the central star is difficult to make out.

Hercules *See page 54.*

Lacerta, the Lizard *See page 61.*

Lynx, the Lynx *See page 49.*

Perseus *See page 67.*

Ursa Major, the Great Bear *See page 94.*

Ursa Minor, the Little Bear
Also called the Little Dipper, this constellation is best known because its brightest star, Polaris, is less than one degree away from the north celestial pole. Polaris, also called the Pole Star and the North Star, is not especially bright, and is best found by using Merak and Dubhe in Ursa Major as pointers.

Polaris is a second-magnitude star and is actually a Cepheid variable, although the change of brightness is not really noticeable. It is also a double star, with a faint (magnitude 9) bluish companion.

Beta, also called Kocab, has about the same brightness as Polaris and is sometimes confused with it, but it is noticeably more orange.

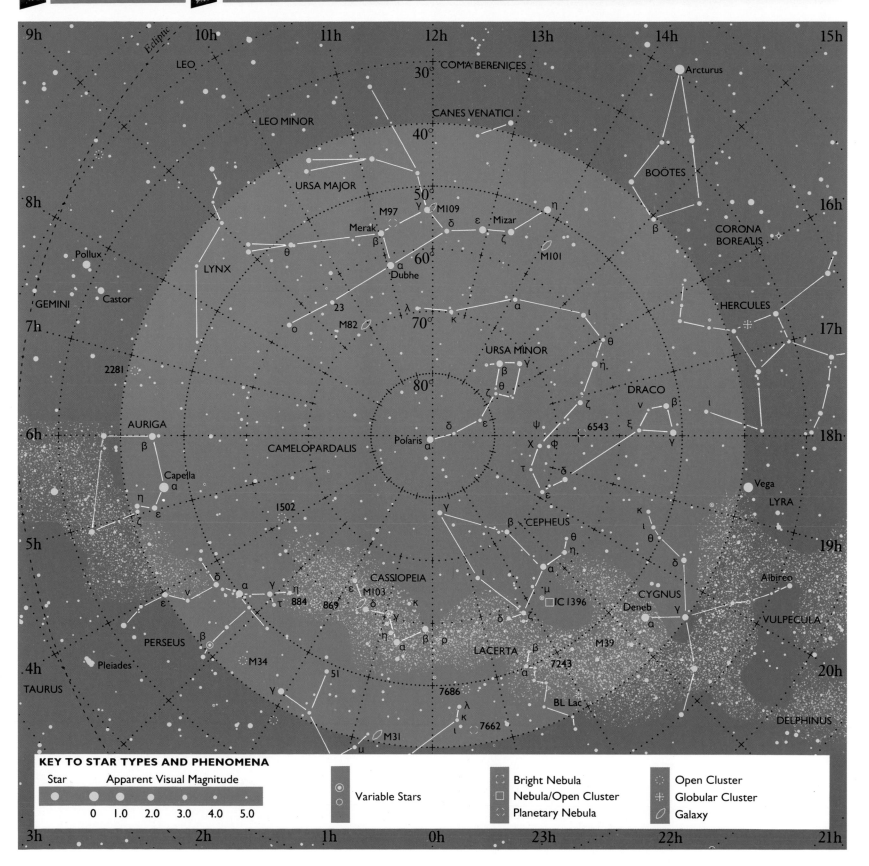

KEY TO STAR TYPES AND PHENOMENA

Star   Apparent Visual Magnitude

0   1.0   2.0   3.0   4.0   5.0

⊙ Variable Stars

Bright Nebula
Nebula/Open Cluster
Planetary Nebula

Open Cluster
Globular Cluster
Galaxy

SOUTH POLAR STARS

The south polar region might lack a convenient pole star, but it is richly endowed with brilliant objects. The Milky Way here is dazzling and contains the unmistakable Southern Cross and its pointers, Alpha and Beta Centauri.

▲ This is the largest of the two Magellanic Clouds that are a feature of far southern skies. Located in Dorado, it is a rather shapeless mass of stars about 30,000 light-years across. The Small Magellanic Cloud, in Tucana, lies about 20,000 light-years farther away and is about 16,000 light-years across.

Ara, the Altar

This constellation lies mainly in the Milky Way and is a good region for sweeping with binoculars. Among its fine globular clusters is NGC6397, one of the closest known. It is found roughly midway between Beta and Theta.

NGC6193, roughly on a line from Theta through Alpha, is the brightest of the open clusters in the constellation and is embedded in nebulosity.

Carina, the Keel

A fine constellation of bright stars, nebulae, and clusters, set in the Milky Way. Its leading star, Canopus, shines like a beacon even in this rich region of the heavens. At magnitude −0.7, it is the second brightest star in the night sky, after Sirius.

From Canopus, Carina extends nearly to Crux. Its stars Iota and Epsilon form, with Delta and Kappa Velae, the so-called False Cross. This is sometimes mistaken for the true Southern Cross of Crux.

Eta is one of the more interesting stars, not particularly bright at present, but once as bright as Canopus. It is embedded in one of the most glorious nebulae in the heavens, the Eta Carinae Nebula (NGC3372), which is visible to the naked eye and splendid in binoculars.

Close to Epsilon is the fine open cluster NGC2516, another naked-eye subject. Bluish-white Theta is sur-

rounded by the bright open cluster IC2602. The open cluster NGC3114 and the globular cluster NGC2808 are among other objects well worth looking at in this stunning constellation.

Centaurus, the Centaur *See page 82.*

Crux, the Southern Cross *See page 82.*

Dorado, the Swordfish

Dorado is a mainly undistinguished constellation. Among the few bright stars, Beta is a Cepheid variable, and one of the brightest known. Dorado is notable mainly because it contains most of the Large Magellanic Cloud (LMC).

The LMC is the nearest galaxy to our own and is clearly visible to the naked eye as a hazy patch of light. It looks beautiful in binoculars or a small telescope, which will resolve some of the brightest stars and nebulae.

One of the nebulae, however, can be seen even with the naked eye. It is the Tarantula Nebula (NGC2070, also called 30 Doradus). It was in this nebula that the spectacular supernova SN1987A exploded in February 1987, the most spectacular for centuries.

Eridanus *See page 66.*

Grus, the Crane *See page 60.*

Lupus, the Wolf *See page 53.*

Puppis, the Poop *See page 47.*

Tucana, the Toucan

Although its stars are quite faint, Tucana has two claims to fame. It includes the Small Magellanic Cloud (SMC) ¨nd one of the two finest globular clusters in the heavens, 47 Tucanae.

The SMC is a neighboring galaxy, a little farther away than the LMC in Dorado and not so spectacular to look at. Nevertheless, even a small telescope will bring into focus nebulae and clusters and even individual stars.

The globular cluster 47 Tucanae (NGC104) lies on the edge of the SMC and vies with Omega Centauri in Centaurus in splendor. It is easily visible to the naked eye and brilliant in binoculars. Another globular cluster, NGC362, lies close by, again right on the edge of the SMC.

Sagittarius, the Archer *See page 88.*

Vela, the Sails *See page 49.*

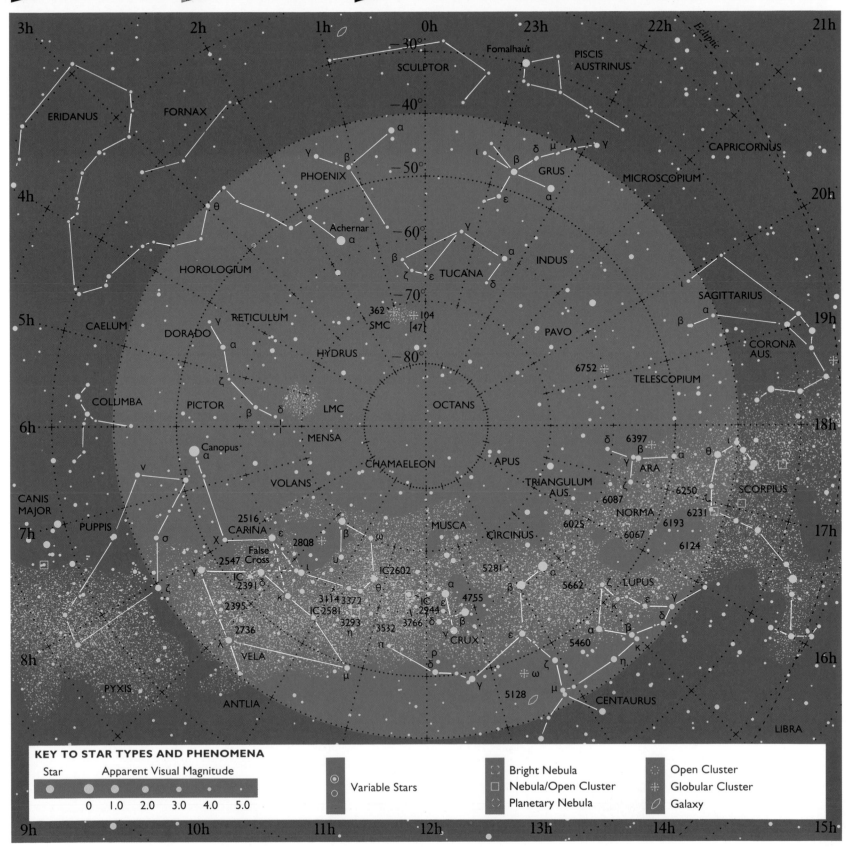

KEY TO STAR TYPES AND PHENOMENA

Star   Apparent Visual Magnitude

0   1.0   2.0   3.0   4.0   5.0

Variable Stars

Bright Nebula
Nebula/Open Cluster
Planetary Nebula

Open Cluster
Globular Cluster
Galaxy

January Stars

*The hexagonal pattern formed by six bright stars –
Sirius, Rigel, Aldebaran, Capella, Pollux, and
Procyon – spans the meridian – the north-south line.
Right on the meridian is a seventh bright star, the
reddish Betelgeux.*

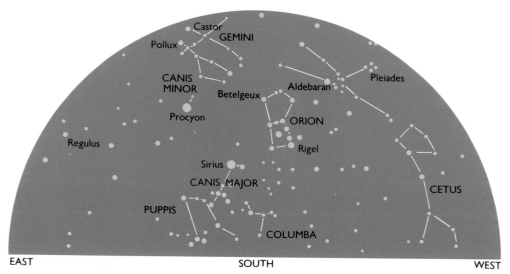

EAST   SOUTH   WEST

▲ **Northern Hemisphere
View of the night sky in
mid-latitudes looking
south at about 11 p.m.
local time on about
January 7.** Orion in mid-sky
acts as a signpost to super-
bright Sirius in the southeast
and to the noticeably orange
Aldebaran in the northwest.

▼ **Southern Hemisphere
View of the night sky in
Australia and southern
Africa looking north at
about 11 p.m. local time
on about January 7.**
In this aspect, the hexagon of
stars appears a perfect shape
and right in the center of the
sky. It is spectacular.

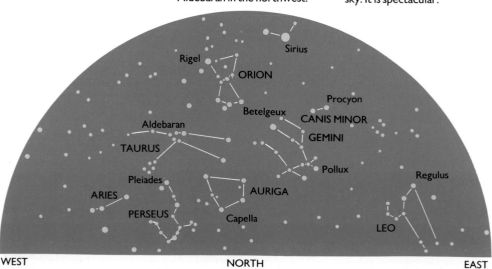

WEST   NORTH   EAST

Auriga, the Charioteer

This conspicuous kite-shaped constellation lies astride
the Milky Way. It can be quickly located by its leading
star, Capella, the sixth brightest star in the sky. Close to
Capella is a triangle of fainter stars, known as the Haedi,
or the Kids. Of the three, Epsilon and Zeta are the most
interesting: they are both eclipsing binaries.

The Epsilon system consists of a yellow supergiant
and a huge dark companion star we cannot see. Usually at
about magnitude 3, it dims to magnitude 4 every 27 years
when the dark companion eclipses the bright star. In the
more colorful Zeta system, the components are an orange
supergiant and a smaller blue star. The change in brightness, every 2 years 8 months, is not easy to detect.

Auriga also contains three bright open clusters, M36,
M37, and M38, easily seen in binoculars. They lie in a
more or less straight line at the edge of the Milky Way
between Theta Aurigae and Beta Tauri in the adjoining
constellation.

Canis Major, the Great Dog

Although only a small constellation, Canis Major boasts
the brightest star in the sky, Sirius (magnitude −1.5),
also called the Dog Star. As stars go, pure white Sirius is
not truly very bright. It appears so bright because it is
relatively close to Earth, less than nine light-years away.
This is scarcely a stone's throw as far as distances in the
universe are concerned.

Sirius is a binary star with a small but dense companion.
This Companion of Sirius (nicknamed appropriately the
Pup) is a white dwarf, and indeed the first of this type of
star to be discovered, in 1862. Although it is magnitude
8, the Pup is difficult to see because it orbits close to
Sirius and tends to be obscured by that star's glare.

Of the other bright stars in Canis Major, the brightest
is Epsilon. It is actually tens of thousands of times more
luminous than Sirius, but lies a hundred times farther away.

Among several open clusters in Canis Major, M41 is
the brightest. Located just south of Sirius, it may in good
viewing conditions be visible to the naked eye as a misty
glow. Small telescopes show up to two dozen bright stars
from the total of about 100.

Another striking open cluster is NGC2362, gathered
around the fourth-magnitude Tau. It contains about 40
very young stars. In binoculars, the whole of this southern
part of the constellation makes spectacular sweeping
since it lies in the Milky Way.

Canis Minor, the Little Dog *See page 46.*

Carina, the Keel *See page 42.*

Columba, the Dove

This constellation has relatively faint stars, but is easily found because it lies in an unusually bare region of the southern skies. It contains a number of globular clusters and galaxies, but only the cluster NGC1851 is visible in small telescopes.

Eridanus *See page 66.*

Gemini, the Twins *See page 46.*

Lepus, the Hare

This constellation lies immediately south of Orion and is thus easy to locate. It is small, but contains some interesting objects. They include R Leporis, found by continuing a line through Alpha and Mu. It is a Mira variable star with a wide variation in brightness, between about magnitudes 5 and 10 over a period of about 14 months. It is one of the reddest stars we know and is sometimes called the Crimson Star.

Gamma is an interesting double star, with orange and yellow components. Equidistant from Gamma and Epsilon and in line with Alpha and Beta is the globular cluster M79. It is small but bright.

Lynx, the Lynx *See page 49.*

Monoceros, the Unicorn *See page 47.*

Orion *See page 86.*

Perseus *See page 67.*

Puppis, the Poop *See page 47.*

Taurus, the Bull *See page 92.*

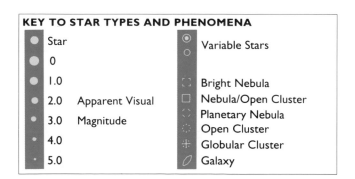

KEY TO STAR TYPES AND PHENOMENA

●	Star	◉	Variable Stars
●	0	○	
●	1.0		
●	2.0	Apparent Visual	⬓	Bright Nebula
●	3.0	Magnitude	☐	Nebula/Open Cluster
●	4.0		⬭	Planetary Nebula
			⬚	Open Cluster
●	5.0		✦	Globular Cluster
			⬭	Galaxy

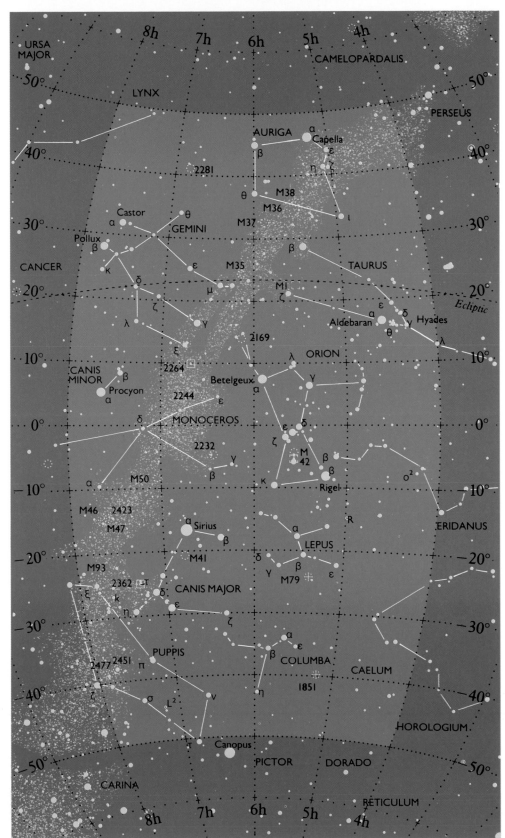

FEBRUARY STARS

The panoply of brilliant stars astride and to the west of the Milky Way has moved from center stage, but is still stunning in the west. However, east of the Milky Way the skies appear rather dull by comparison.

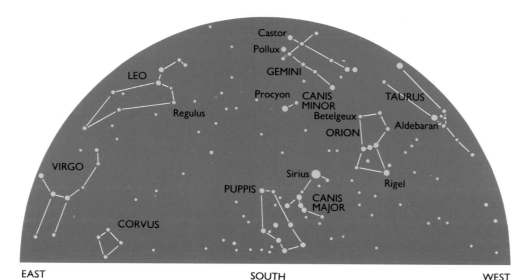

EAST   SOUTH   WEST

▲ **Northern Hemisphere View of the night sky in mid-latitudes looking south at about 11 p.m. local time on about February 7.** The bright stars in mid-sky line up more or less parallel with the horizon. From the east, they are Regulus, Procyon, Betelgeux, and Aldebaran.

▼ **Southern Hemisphere View of the night sky in Australia and southern Africa looking north at about 11 p.m. local time on about February 7.** The hexagonal arrangement of bright stars has slipped west, with Capella barely visible. In the east, Leo is ascending.

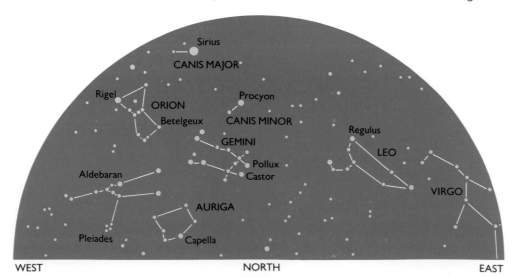

WEST   NORTH   EAST

Auriga, the Charioteer *See page 44.*

Cancer, the Crab
Sandwiched between Leo and Gemini, Cancer is one of the constellations of the zodiac. Though relatively faint, it contains much of interest. In the middle of the constellation is the beautiful open cluster M44, called Praesepe, or the Beehive. It is one of the most prominent clusters, easily visible to the naked eye. Binoculars show up dozens of its stars; while large telescopes reveal hundreds.

Close to Alpha is another open cluster, M67, visible in binoculars, but needing a telescope to bring out individual stars. Cancer also features two lovely double stars, Iota and Zeta. Both have yellow and blue components.

Canis Major, the Great Dog *See page 44.*

Canis Minor, the Little Dog
Little by name and little in size, this constellation is notable for its one bright star, Procyon. At a distance of about 11 light-years, Procyon is the third closest bright star to Earth, after Alpha Centauri and Sirius. And like Sirius, it has a faint white dwarf companion.

Carina, the Keel *See page 42.*

Columba, the Dove *See page 45.*

Gemini, the Twins
This splendid constellation of the zodiac is well named, for it boasts twin first-magnitude stars, Castor and Pollux. Pollux is the brighter of the two and is more colorful, being a rich yellow-orange.

Castor, however, is the more interesting twin because it is a multiple star system. A small telescope resolves Castor into two components and brings a fainter third into view. The spectra of the three stars reveal that each is a spectroscopic binary, with a close companion, making a six-star system in all.

Among Gemini's double stars is Zeta. Its brightest component is a Cepheid variable, which fluctuates in brightness over a period of ten days. The change in brightness can be seen by comparison with the steady shining Delta, which a small telescope will reveal to be a double.

A good binocular subject in the constellation is M35, a large and bright open cluster just north of Mu and Eta. Telescopes will reveal close to Delta the greenish disk of NGC2392, a fine planetary nebula.

Hydra, the Water Snake *See pages 48 and 50.*

Leo, the Lion *See page 48.*

Lepus, the Hare *See page 45.*

Lynx, the Lynx *See page 49.*

Monoceros, the Unicorn

A faint constellation, Monoceros is located between the two "Dogs," Canis Minor and Major. It has no bright stars, but spans a particularly rich area of the Milky Way.

Beta is one of the most interesting stars. It is a multiple star, whose three components can be seen as a neat triangle even in small telescopes.

Close by Epsilon is a striking open cluster, NGC2244. Large telescopes show it to be surrounded by one of the most beautiful nebulae in the heavens, the flower-like Rosette Nebula (NGC2237). Another nebula, the Cone Nebula, surrounds the open cluster NGC2264, just south of Chi Geminorum. Midway between Alpha and Beta is another open cluster, M50.

Puppis, the Poop

Puppis is a fine southern constellation nestling in the Milky Way. It abounds in rich star fields and clusters. The second-magnitude Zeta is the brightest star. Electric blue in color, it is one of the hottest stars known, with a surface temperature of tens of thousands of degrees.

Numerous doubles dotted around the constellation include Kappa and Sigma. Pi is one of a quartet of stars, well seen in a small telescope. Close to Sigma is L^2, a long-period variable that changes noticeably in brightness from magnitudes 3 to 6 over a period of about five months.

Among the many open clusters are the naked-eye M47 and the fainter M46, which are located side by side and close to NGC2423. M93 is farther south, near Xi. Close to Zeta is NGC2477, containing hundreds of faint stars so densely packed that it almost looks like a globular cluster.

Ursa Major, the Great Bear *See page 94.*

Vela, the Sails *See page 49.*

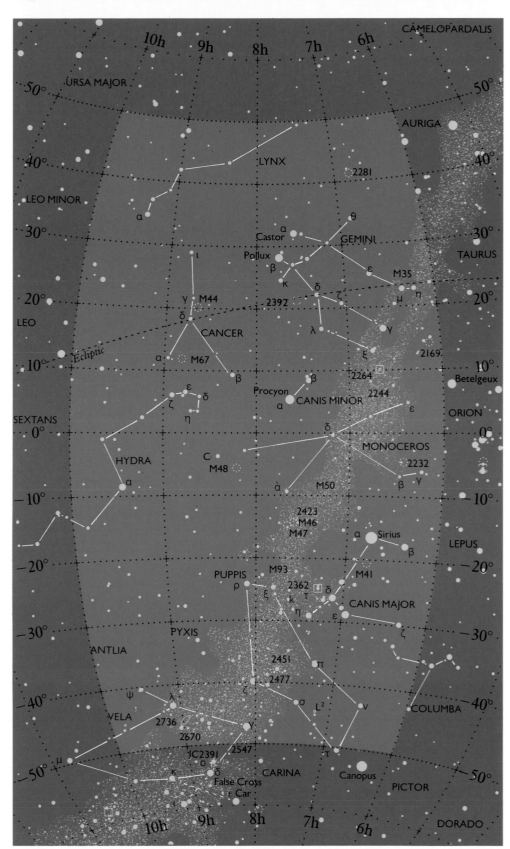

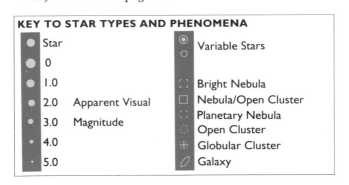

KEY TO STAR TYPES AND PHENOMENA

Star		Variable Stars
0		
1.0		Bright Nebula
2.0	Apparent Visual	Nebula/Open Cluster
3.0	Magnitude	Planetary Nebula
		Open Cluster
4.0		Globular Cluster
5.0		Galaxy

MARCH STARS

While Orion takes its curtain call in the west, the majestic Leo, with its distinctive "Sickle" head, occupies the mid-sky. Its presence heralds the onset of spring in the Northern Hemisphere and of fall in the Southern.

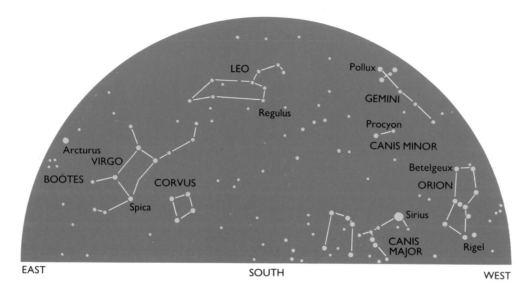

EAST   SOUTH   WEST

▲ **Northern Hemisphere View of the night sky in mid-latitudes looking south at about 11 p.m. local time on about March 7.** Rigel is close to setting in the west. Sirius is also low down and about to make its exit. In the east, Spica is making its entrance.

▼ **Southern Hemisphere View of the night sky in Australia and southern Africa looking north at about 11 p.m. local time on about March 7.** Regulus appears due north in the mid-sky. Spica and Arcturus appear in the east, the one vertically above the other. other.

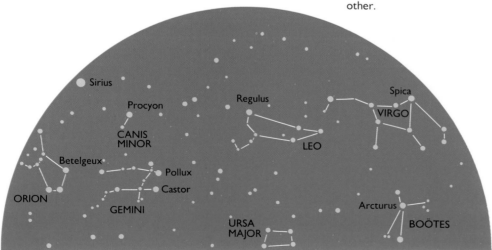

WEST   NORTH   EAST

Cancer, the Crab *See page 46.*

Gemini, the Twins *See page 46.*

Hydra, the Water Snake (head)
Hydra winds its way, serpent-like, more than a quarter of the way around the celestial sphere. Indeed, it is the largest of all the constellations. It rears its head near Canis Minor, just north of the celestial equator, while its tail extends as far as Libra.

The only bright star in Hydra, the second-magnitude Alpha, is called Alphard, meaning the "Solitary One" because it is the only reasonably bright star in a barren region of the heavens. It is also known as Cor Hydrae, or the Hydra's Heart.

The stars in the Hydra's head make an attractive group in low-powered binoculars. One of them, Epsilon, is a colorful double, separated in larger telescopes. M48 is a bright cluster approaching naked-eye visibility, found by moving southward from the head past the close trio of stars, c.

The brightest planetary nebula in the heavens, NGC3242, lies in Hydra, just south of Mu. Small telescopes show this magnitude 9 object as a faint bluish-green disk. Sometimes called the Ghost of Jupiter Nebula, its central star is clearly seen. (For Hydra's tail, see page 50.)

Leo, the Lion
This constellation of the zodiac is one that looks passably like the animal it is supposed to represent. The most recognizable shape in Leo is the Sickle, the sickle-like arrangement of stars that forms the lion's head and mane.

In the Sickle, Gamma, Mu, and Lambda are all orange in color. Gamma, often called Algeiba, the Lion's Mane, is a double, which can be resolved in a telescope into a pair of golden yellow stars.

Brighter than the Sickle stars is the first-magnitude Regulus, also called Cor Leonis, or the Lion's Heart. It is a double, with the fainter star of the eighth magnitude.

The other prominent shape in Leo is at the tail end, where there is a triangle of bright stars. Of this trio Beta, also called Denebola, is the brightest at about magnitude 2.

Beginning just south of Delta is a region rich in galaxies. The brightest ones lie south of a line joining Beta and Regulus. South of Theta are M65 and M66, both about magnitude 9 and both visible in binoculars in good seeing conditions. Like our own Galaxy, they are spirals, although large telescopes are needed to bring out their spiral form.

About halfway between them and Regulus is another pair of spiral galaxies, M95 and M96. All four galaxies lie

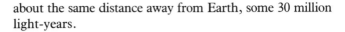

about the same distance away from Earth, some 30 million light-years.

Lynx, the Lynx

Said to be so named because you have to be lynx-eyed to spot it, this is a faint constellation in a relatively empty part of the northern heavens. The third-magnitude Alpha is distinctly red.

Monoceros, the Unicorn *See page 47.*

Puppis, the Poop *See page 47.*

Ursa Major, the Great Bear *See page 94.*

Vela, the Sails

Vela spans the Milky Way in the southern skies and, like the adjacent constellations Centaurus, Puppis, and Carina, is rich in star fields and clusters. In the constellation, the Milky Way splits up because of distant obscuring dust clouds.

Kappa and Delta are second-magnitude stars which, along with Iota and Epsilon in Carina, form a noticeable cross-shape, the so-called False Cross. The False Cross has much the same orientation as the true cross of Crux close by, but is larger.

Immediately north of Delta is Omicron, which forms part of the cluster IC2391. Delta, Omicron, and the rest of the cluster look great through binoculars. Another good binocular subject is NGC2547, just south of Gamma.

Gamma is easy to spot as a double, and its blue-white brighter component is particularly hot and luminous. It is a type known as a Wolf-Rayet star, an unstable body thought to be on the brink of becoming a supernova and blasting itself apart.

South of Lambda, long-exposure photographs reveal wisps of an extensive nebula, known as the Gum Nebula (NGC2736). It is a supernova remnant, the gaseous remains of a supernova explosion that took place about 11,000 years ago.

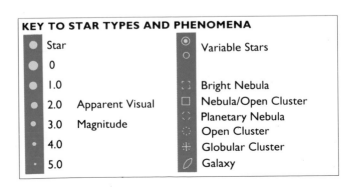

KEY TO STAR TYPES AND PHENOMENA

● Star	◉	Variable Stars
● 0	○	
● 1.0	⬒	Bright Nebula
● 2.0 Apparent Visual	⬜	Nebula/Open Cluster
● 3.0 Magnitude	◇	Planetary Nebula
	⬡	Open Cluster
● 4.0	✵	Globular Cluster
● 5.0	⬭	Galaxy

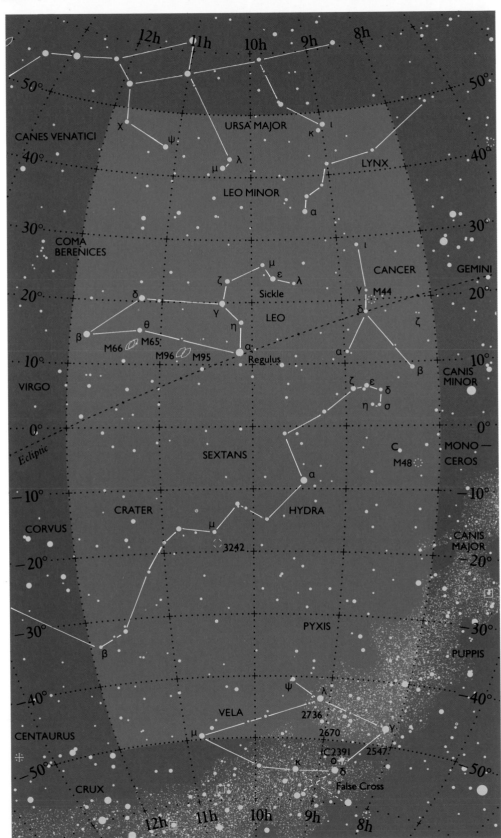

April Stars

Leo is still prominent, but as a whole the sky looks bare because of the presence of the sprawling Hydra and Virgo. These, the two largest constellations in the heavens, have no very bright stars, except for Spica.

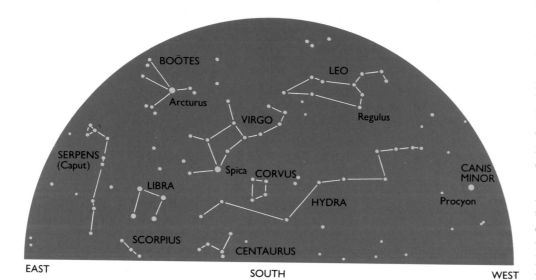

EAST   SOUTH   WEST

▲ **Northern Hemisphere View of the night sky in mid-latitudes looking south at about 11 p.m. local time on about April 7.** High in the east, Arcturus is the brightest star, forming an expansive triangle with Regulus and Spica. Procyon lies in the west.

▼ **Southern Hemisphere View of the night sky in Australia and southern Africa looking north at about 11 p.m. local time on about April 7.** This is a good month to experience the extragalactic delights of the "empty" region between Leo and Boötes.

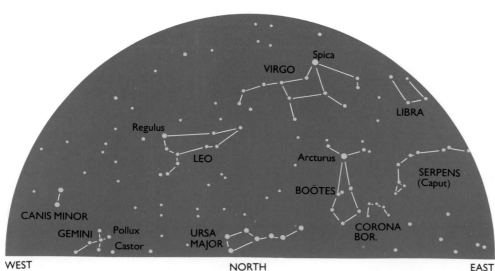

WEST   NORTH   EAST

Canes Venatici, the Hunting Dogs

Overall, this is a faint constellation, located underneath the handle of the Big Dipper. However, it boasts one of the finest globular clusters in northern skies and one of the most distinctive galaxies.

Even its leading star, Alpha, is only slightly brighter than the third magnitude. It is also called Cor Caroli, meaning Charles's Heart, a name given to it by Britain's second Astronomer Royal, Edmond Halley. The heart in question is that of the executed Charles I of England. Alpha is a double, which a small telescope will resolve.

The first of the telescopic highlights of Canes Venatici is the M3 globular cluster. It is located about halfway between Cor Caroli and the bright Arcturus in the neighboring constellation Boötes. At about magnitude 6, it is on the limit of naked-eye visibility, and binoculars show it as a fuzzy patch. Telescopes will resolve it into an enormous globe of closely packed stars.

The other telescopic highlight of the constellation is M51, the famous Whirlpool Galaxy. Small telescopes show it as a pair of hazy spots. Larger instruments bring out its classic face-on spiral shape. The two spots in fact turn out to be the centers of two connected galaxies. Historically, the Whirlpool is important because it was the first galaxy to have its spiral structure resolved.

Like the neighboring constellation Coma Berenices, Canes Venatici is richly endowed with galaxies. Among the brighter ones worth looking for are M63, M94, and M106. M63 is located just north of 20 Canes Venatici; M94, just over halfway from 20 to Beta; and M106, from Beta about halfway to Gamma Ursa Majoris. Larger telescopes will reveal a string of fainter galaxies going south from M106.

Centaurus, the Centaur *See page 82.*

Coma Berenices, Berenice's Hair *See page 52.*

Corvus, the Crow

A tiny constellation on Hydra's back, easy to spot because of its quartet of leading stars, all of (or slightly below) third magnitude. They make a nice group in binoculars. Delta is an easy double, with components of contrasting colors.

Hydra, the Water Snake (tail)

Hydra continues to wind through this segment of the celestial sphere. Among viewable objects in this, the tail end of the water snake, is the star R, located near Gamma. It is a Mira variable star, distinctly red. It varies markedly in brightness between magnitudes 4 and 11 over a period of just over a year.

Due south of R, on the edge of the constellation, is one of the brightest face-on spiral galaxies in the heavens, M83. Larger instruments reveal it is a barred spiral galaxy, with a distinct bar across the middle.

Leo, the Lion *See page 48.*

Ursa Major, the Great Bear *See page 94.*

Vela, the Sails *See page 49.*

Virgo, the Virgin *See page 96.*

▼ One of the most distinctive objects in the heavens, the Whirlpool Galaxy (M51). It is actually two connected galaxies. The main spiral is noted for its bright nucleus and well-defined arms.

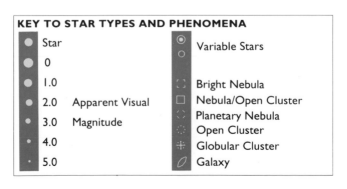

KEY TO STAR TYPES AND PHENOMENA

	Star		Variable Stars
	0		
	1.0		
	2.0 Apparent Visual		Bright Nebula
	3.0 Magnitude		Nebula/Open Cluster
			Planetary Nebula
	4.0		Open Cluster
	5.0		Globular Cluster
			Galaxy

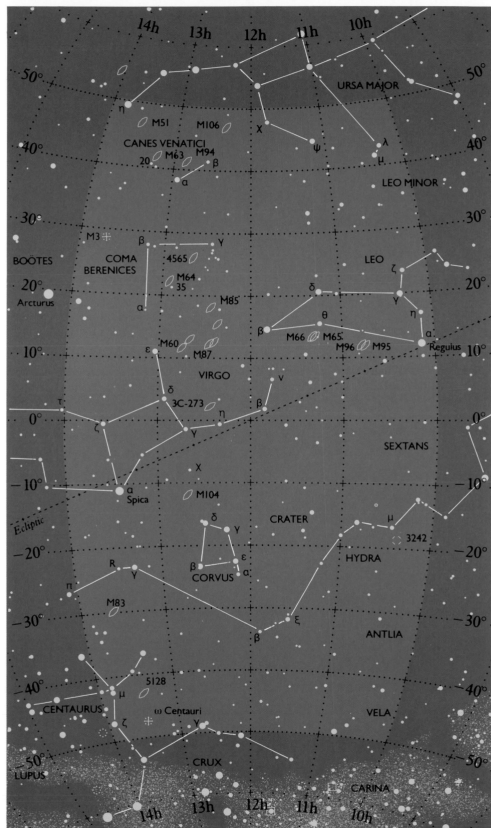

MAY STARS

Arcturus and Spica are close to the meridian this month and provide a pleasing color contrast. Arcturus is a lovely orange, while Spica is pure white. Both are giants, the one cool, the other very hot.

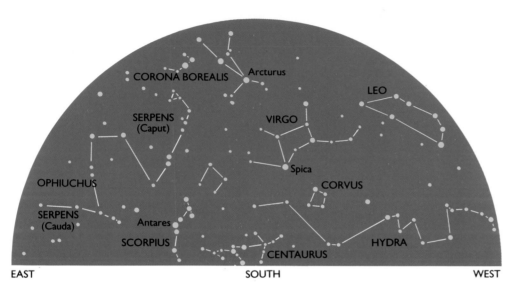

EAST   SOUTH   WEST

▲ **Northern Hemisphere View of the night sky in mid-latitudes looking south at about 11 p.m. local time on about May 7.**
Colored Arcturus has a rival appearing over the horizon, red Antares. The whole of Serpens – Caput and Cauda – is now visible in the east.

▼ **Southern Hemisphere View of the night sky in Australia and southern Africa looking north at about 11 p.m. local time on about May 7.**
Leo is slipping toward the horizon. At lower latitudes, the Big Dipper may just be glimpsed on the horizon.

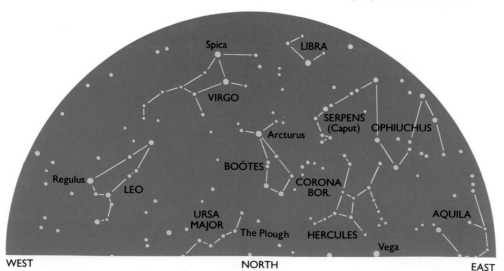

WEST   NORTH   EAST

Boötes, the Herdsman

This fine kite-shaped constellation boasts the brightest star in northern skies and the fourth brightest star in the whole heavens. This star is Arcturus ("Bear-keeper"). It is a red giant, which explains its noticeable orange color. It can perhaps best be found by following around the curve of the handle of the Big Dipper.

There are several doubles of note in Boötes, none finer than Epsilon. The star appears only slightly colored in binoculars, but a telescope separates it into beautiful yellowish-orange and bluish-green components, making it one of the most attractive doubles in the heavens.

Xi is another colorful double, which even small telescopes should be able to separate into a yellow and orange pair. The third fine double in the constellation is Mu. Binoculars will separate it into two yellow stars. A telescope will split the fainter of the two into a further yellow pair.

Boötes has few interesting deep-sky objects, such as galaxies and clusters. There is, however, one reasonable globular cluster, NGC5466, which forms one corner of a parallelogram with Rho, Alpha, and Eta.

Canes Venatici, the Hunting Dogs *See page 50.*

Centaurus, the Centaur *See page 82.*

Coma Berenices, Berenice's Hair

Like Canes Venatici to the north, Coma Berenices is faint and inconspicuous at first sight, but look at it in binoculars, small, and large telescopes, and it becomes increasingly richer with every increase in magnification.

Essentially, it is a constellation dominated by galaxies. Literally hundreds of them lie in this region of the heavens, forming the so-called Coma Cluster. At a distance of some 450 million light-years, these galaxies are mostly beyond the range of amateur telescopes. However, there are quite a few much closer galaxies, which are visible in small telescopes. M64 is one of them. It is located on a line joining Alpha and Gamma, near star 35. Large telescopes show it as an ellipse, with a distinctive dark spot. For this reason, it is known as the Black-Eye Galaxy.

Farther along the Alpha-to-Gamma line is the slightly fainter NGC4565, known as the Needle Galaxy. It appears needle-like because we are looking at it edge-on. The galaxies M85 and M100, both of the ninth magnitude, form part of a dense grouping of galaxies at the southern edge of the constellation, which continues into Virgo. This whole region is a fascinating one for the telescopic browser.

Corona Borealis, the Northern Crown *See page 54.*

Corvus, the Crow *See page 50.*

Hydra, the Water Snake *See pages 48 and 50.*

Libra, the Scales

Libra is a faint constellation of the zodiac, which lacks any really bright stars. It perhaps suffers from being cheek by jowl with the bright stars in the head of Scorpius.

Indeed, originally, the stars in Libra formed the claws of the Scorpion. Its leading star Alpha (magnitude 2.7) is named Zubenelgenubi, meaning the "southern claw." It is slightly less bright than Beta, the "northern claw" Zubenelchemale.

Alpha is a fine wide double star, which binoculars separates easily. If the dim component (fifth magnitude) were somewhat brighter, Alpha would just be separable with the naked eye.

Beta has an unusual greenish color. Delta is an eclipsing binary, similar to Algol. Every 56 hours, the dim component passes in front of the bright one, and the brightness of the system dips briefly from the fourth to the sixth magnitude.

Lupus, the Wolf

This constellation is sandwiched between Scorpius and Centaurus on the edge of the Milky Way. It is therefore a rewarding region for sweeping with binoculars. There are several doubles worth looking at in the constellation, including Pi, Kappa, and Eta.

As for clusters, there are many that can be found in larger telescopes. More modest instruments will be able to pick up the globular cluster NGC5986, close to Eta, and the open cluster NGC5822, close to Zeta.

Scorpius, the Scorpion *See page 90.*

Serpens Caput, the Serpent's Head *See page 55.*

Ursa Major, the Great Bear *See page 94.*

Virgo, the Virgin *See page 96.*

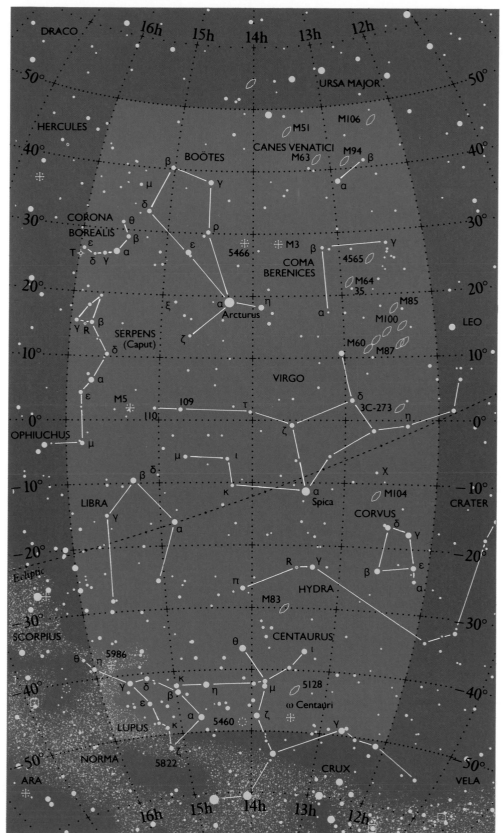

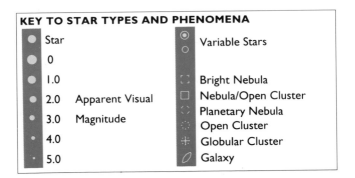

KEY TO STAR TYPES AND PHENOMENA

Star			Variable Stars
0			
1.0			Bright Nebula
2.0	Apparent Visual		Nebula/Open Cluster
3.0	Magnitude		Planetary Nebula
			Open Cluster
4.0			Globular Cluster
5.0			Galaxy

June Stars

In this month and July, Scorpius is a constellation to watch, low in northern skies and near the zenith in southern. Its brightest star, Antares, joins Arcturus, Vega, Deneb, Altair, and Spica in the night sky starscape.

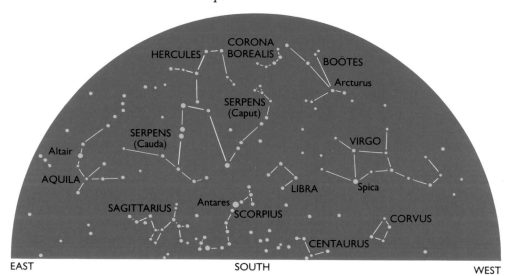

▲ Northern Hemisphere View of the night sky in mid-latitudes looking south at about 11 p.m. local time on about June 7. The heads of the Serpent and the Scorpion lie due south. Compare the colors of Antares low down and Arcturus high up.

▼ Southern Hemisphere View of the night sky in Australia and southern Africa looking north at about 11 p.m. local time on about June 7. This month provides a good opportunity for observing the northern constellations Boötes and Hercules.

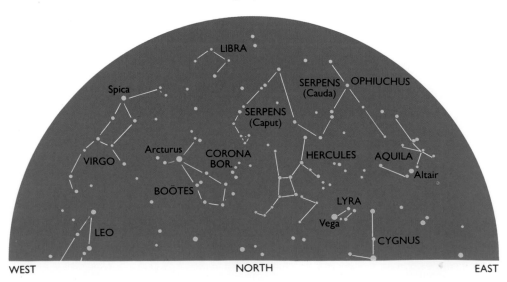

Ara, the Altar *See page 42.*

Boötes, the Herdsman *See page 52.*

Centaurus, the Centaur *See page 82.*

Corona Borealis, the Northern Crown
This is a tiny, but easily recognizable constellation, a neat semicircle of stars with the second-magnitude Alpha in the middle of the arc. It contains a number of doubles, including Zeta and Nu, but is particularly notable for two remarkable variable stars.

One is R, located north of Delta in the "bowl." For most of the time, R is about magnitude 6 and readily visible in binoculars, but from time to time it suddenly dims to about magnitude 12 or even lower. It may recover its brightness in a matter of months, but sometimes takes years. Astronomers think one explanation of the sudden dimming might be that the star periodically ejects clouds of sooty material, which blocks its light.

The other fascinating variable is T, located south of Epsilon. This star is usually only about magnitude 10, but sometimes flares up to brighter than 6 and becomes visible to the naked eye. Indeed, in 1866 and 1946, T became as bright as the second magnitude. Aptly called the Blaze Star, T periodically flares up in this way, over a matter of hours, because it is a type called a recurrent nova. No one knows when it may flare up again, so it is always worth watching.

Underneath the "bowl," beyond Beta, large telescopes will show one of the densest clusters of galaxies known, called Abell 2065. Over 400 galaxies are clustered closely together about 1,000 million light-years away.

Draco, the Dragon *See page 40.*

Hercules
Hercules is a large constellation, in fact, the fifth largest in the heavens. It is not overly impressive to the naked eye, but is richly endowed to the binocular and telescopic observer.

Its pièce de résistance is M13, the finest globular cluster in northern skies, just about visible to the naked eye on a really dark night. M13 is found on a line joining Zeta and Eta. Binoculars show it as a fuzzy blob, but telescopes resolve it into a conglomeration of countless stars.

M92 is another fine globular cluster, slightly fainter than M13, found on a line joining Eta with Iota. In larger telescopes NGC6210 is well worth observing. It is a small, bright planetary nebula with a brilliant blue central star.

Among the stars in Hercules, Alpha is noticeably red

and varies over a period of months between the third and fourth magnitudes. It is also a double star with reddish and greenish components. Delta, Rho, and Mu are also doubles worth investigating.

Hydra, the Water Snake (tail) *See page 50.*

Libra, the Scales *See page 53.*

Lupus, the Wolf *See page 53.*

Lyra, the Lyre *See page 56.*

Ophiuchus, the Serpent-Bearer *See page 56.*

Sagittarius *See page 88.*

Scorpius, the Scorpion *See page 90.*

Serpens Caput, the Serpent's Head

This is the head part of Serpens, the Serpent, a constellation that is split into two by Ophiuchus, the Serpent-Bearer. The tail part is Serpens Cauda (see page 57). The triangle of stars in the Head look attractive in the same field in binoculars. Kappa and Rho to the north are distinctly reddish.

Also in the same field, close to Beta, is R, a Mira variable with a wide variation in brightness. At maximum, it reaches magnitude 6, but at minimum fades beyond the reach of binoculars. Its period is slightly less than a year. Delta and the orange Alpha are visible as doubles in a telescope.

However, the Head's most impressive object is the globular cluster M5, one of the finest of its type in northern skies. You can find it, close to the dim star 5, by continuing a line through Lambda and Alpha. It is also in line with the stars 110 and 109 Virginis in the neighbouring constellation Virgo.

Virgo, the Virgin *See page 96.*

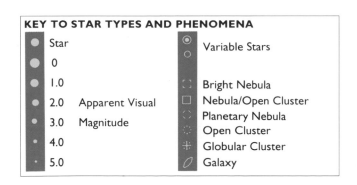

KEY TO STAR TYPES AND PHENOMENA

⬤ Star	◉	Variable Stars
⬤ 0	○	
⬤ 1.0	⊡	Bright Nebula
⬤ 2.0 Apparent Visual	▢	Nebula/Open Cluster
• 3.0 Magnitude	⬦	Planetary Nebula
• 4.0	⊙	Open Cluster
• 5.0	✲	Globular Cluster
	⬭	Galaxy

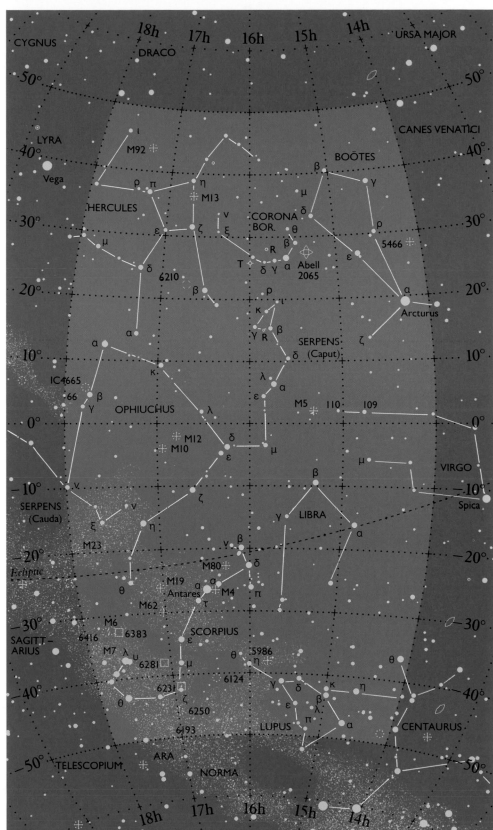

JULY STARS

Vega is high in northern skies and low in southern. With Altair and Deneb, it makes up the Summer Triangle. The Milky Way, particularly the Sagittarius region, is now well placed for viewing in both hemispheres.

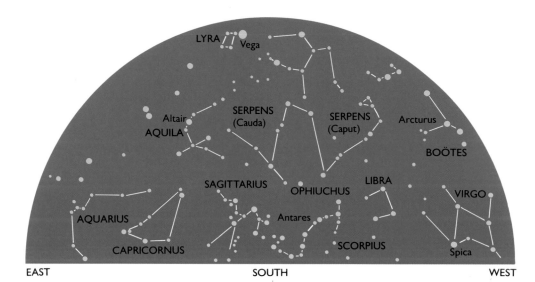

EAST   SOUTH   WEST

▲ Northern Hemisphere View of the night sky in mid-latitudes looking south at about 11 p.m. local time on about July 7. Sagittarius joins Scorpius near the southern horizon, offering northern observers a tantalizing glimpse of the delights of southern skies.

▼ Southern Hemisphere View of the night sky in Australia and southern Africa looking north at about 11 p.m. local time on about July 7. The Swan and the Eagle take wing along the Milky Way. The stars Deneb and Altair lie on a near-vertical line.

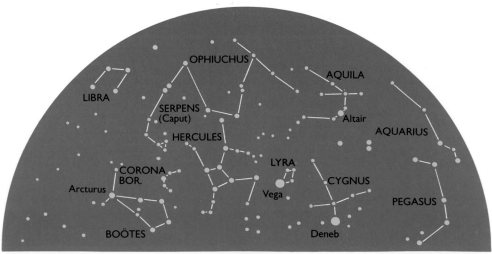

WEST   NORTH   EAST

Aquila, the Eagle *See page 58.*

Cygnus, the Swan *See page 84.*

Hercules *See page 54.*

Lyra, the Lyre

Although this is only a small constellation, it is packed with interesting objects, particularly a surfeit of fine double stars.

The only really bright star in Lyra is the first-magnitude Vega, sometimes called the Harp Star. The fifth brightest star in the heavens, it shines brilliantly in summer skies in the Northern Hemisphere and is one of the trio of stars (along with Deneb in Cygnus and Altair in Aquila) that make up the Summer Triangle.

Close to Vega lies Epsilon, the well-known "double-double" star. It is so-called because it is a double star easily visible in binoculars. And when its two components are observed in a small telescope, they are also seen to be double.

Immediately south of Vega, four stars form a parallelogram. Astronomically, the most interesting of them is Beta. First, Beta is a double star, with yellowish and bluish components. In addition, the brighter yellowish star is an eclipsing binary system. Its two stars are very close together, probably less than 22,000,000 miles (35,000,000 kilometers). Because of this, they are almost certainly egg-shaped, with gas streaming between them.

Beta changes from the third to the fourth magnitude every 13 days. And its variation can readily be followed using Delta or Zeta (both a steady magnitude 4.3) for comparison. Both these stars are also doubles, the Delta system easily being visible in binoculars.

Sandwiched between Beta and Gamma is one of the most striking of all heavenly objects, M57, the Ring Nebula, a colorful "smoke ring" puffed out by a dying star. It is not especially impressive in small telescopes and requires long exposures in larger instruments to show its true splendor.

Ophiuchus, the Serpent-Bearer

This is a large constellation, which splits Serpens, the Serpent, into two – Caput (Head) and Cauda (Tail). In the south, it enters a particularly rich region of the Milky Way, where globular clusters abound.

Among these southern clusters is M19, found midway between Theta and Antares in the adjacent constellation Scorpius. Just north of Antares, on the Ophiuchus/Scorpius boundary, is the faint star Rho. It is worth scanning this region because it is occupied by extensive clouds of gas and dust. Even in binoculars you may be

able to make out the dust lanes.

Two bright globular clusters, M10 and M12, lie in the relatively barren interior of Ophiuchus. M12 makes a triangle with Lambda and Delta, and M10 is close by.

Among the stars, 70 is a colorful orange and yellow double visible in telescopes. Close to star 66 is a faint (magnitude 9) red star called Barnard's Star. It is interesting for two related reasons. First, it is the nearest star to Earth after Proxima and Alpha Centauri, only 5.9 light-years away. Second, it is the star with the largest proper motion, or visible motion across the sky: over 10 seconds of arc per year.

Sagitta, the Arrow *See page 59.*

Sagittarius, the Archer *See page 88.*

Scorpius, the Scorpion *See page 90.*

Scutum, the Shield

Although it is a small constellation with no really bright stars, Scutum lies on the edge of a rich part of the Milky Way. It abounds in star clouds and dark nebulous regions.

The constellation's two highlights lie close together. Just south of Beta is R, a variable star that changes brightness irregularly between about the sixth and ninth magnitudes. Close by is a beautiful open cluster, M11, called the Wild Duck because its fan of stars is said to look like a duck in flight.

Serpens Cauda, the Serpent's Tail

This is the tail part of Serpens, a constellation split by Ophiuchus. The head part is Serpens Caput (*see page 55*). Perhaps the most interesting object in Serpens Cauda is M16, a scattered cluster of stars visible in binoculars quite close to Gamma Scuti in the adjacent constellation. Telescopes show the stars embedded in a glowing nebula, called the Eagle Nebula.

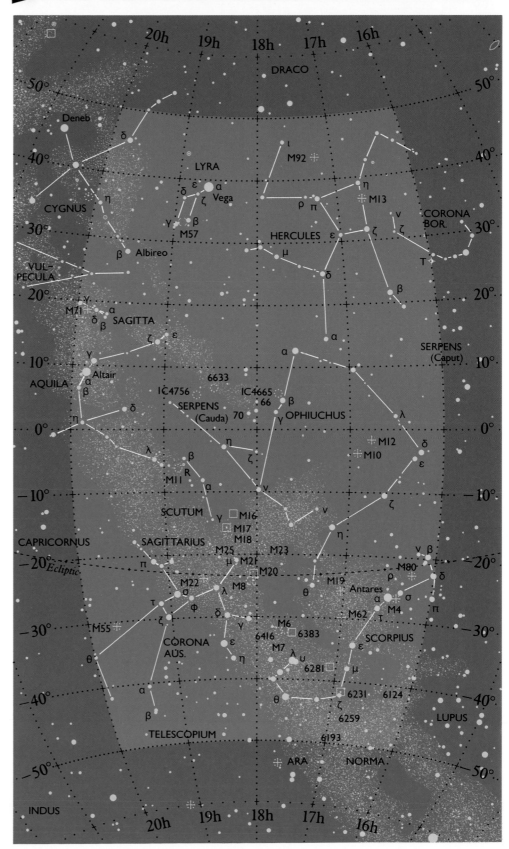

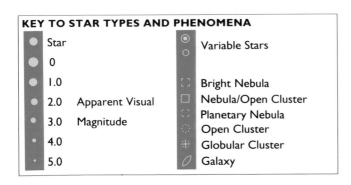

KEY TO STAR TYPES AND PHENOMENA

●	Star	◉	Variable Stars
●	0	○	
●	1.0		
●	2.0 Apparent Visual	⌞⌝	Bright Nebula
●	3.0 Magnitude	▢	Nebula/Open Cluster
·	4.0	◇	Planetary Nebula
·	5.0	⁙	Open Cluster
		✢	Globular Cluster
		⬭	Galaxy

August Stars

The Summer Triangle of the brilliant white Deneb, Vega, and Altair spans the meridian, high in the sky in the Northern Hemisphere, low in the Southern. The appearance of Pegasus in the east, however, hints at a changing season.

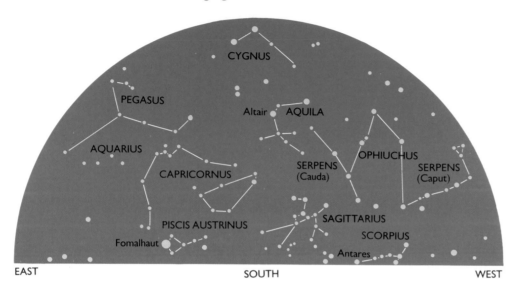

▲ Northern Hemisphere View of the night sky in mid-latitudes looking south at about 11 p.m. local time on about August 7.
Part of Sagittarius is still visible this month; so, at lower latitudes, is Antares. But at higher latitudes, this star is just below the horizon.

▼ Southern Hemisphere View of the night sky in Australia and southern Africa looking north at about 11 p.m. local time on about August 7.
Vega and Deneb are bright and prominent low down in the south, while Altair appears higher up.

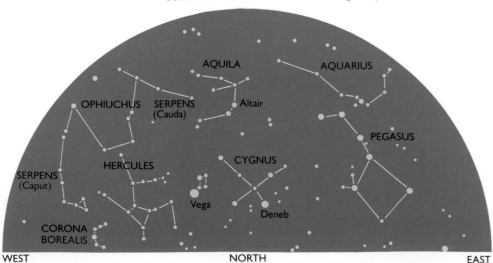

Aquarius, the Water-Bearer *See page 60.*

Aquila, the Eagle
This is a fine constellation, whose curve of bright stars does rather resemble the wings of a bird in flight. Aquila straddles the Milky Way and the celestial equator, and contains rich star fields that are a delight to sweep with binoculars.

The brightest star, Altair, is one of the trio that form the prominent Summer Triangle. (The others are Deneb in Cygnus and Vega in Lyra.) Altair is a brilliant pure white star, contrasting nicely with one of its neighbors, the fainter Gamma, which is distinctly orange.

South of Altair is a noticeable line of three stars. Theta and Delta flank the central Eta, which is a Cepheid variable. It varies between the third and fourth magnitudes over about a week. Its progress can be followed by reference to the third-magnitude Theta and Delta and the fourth-magnitude Iota.

The curve of stars around Lambda acts as a good pointer to the Wild Duck cluster M11, in the neighboring constellation Scutum.

Capricornus, the Sea Goat
Capricornus is one of the constellations of the zodiac, but is fairly inconspicuous. Its brightest stars, of about the third magnitude, form a pattern rather like a crooked triangle.

Alpha is a naked-eye double, but the two stars are not physically associated: the fainter one is more than ten times farther away than the brighter one. Both stars are themselves doubles. Beta is also double and is a true binary star, whose components travel through space together.

The only noteworthy cluster in the constellation is M30, just outside the "crooked triangle" near Zeta. It is a globular cluster, visible in small telescopes, but needing larger ones to resolve the stars properly.

Cygnus, the Swan *See page 84.*

Delphinus, the Dolphin
You can easily imagine this tiny constellation to be a leaping dolphin. Its bright stars form a group so compact that they almost look like an open cluster. In binoculars they all appear in the same field. In the main quadrilateral of stars, Gamma is a fine double, separated in small telescopes into golden yellow components.

Grus, the Crane *See page 60.*

Lacerta, the Lizard *See page 61.*

Lyra, the Lyre *See page 56.*

Sagitta, the Arrow

This is one of the smallest constellations in the heavens, but it has quite a distinctive, arrow-like shape. It lies within the Milky Way, almost due north of the bright stars in Aquila.

To the naked eye, Sagitta appears a rather bland region of the sky, but in binoculars it is well worth viewing for the beautiful star fields it contains. The four main stars are also an attractive sight in binoculars. About midway between Delta and Gamma, small telescopes will show a bright, dense grouping of stars, M71, which is almost certainly a globular rather than an open cluster.

Sagittarius, the Archer *See page 88.*

Scutum, the Shield *See page 57.*

Serpens Cauda, the Serpent's Tail *See page 57.*

Vulpecula, the Fox

Located south of Cygnus, this constellation is ill-defined and devoid of really bright stars. Binoculars and small telescopes, however, show some attractive star fields in this filamentary region of the Milky Way, as well as a number of other interesting objects.

Perhaps the finest is the Dumbbell Nebula, M27, one of the brightest planetary nebulae in the heavens. It is located due north of Gamma Sagittae in the neighboring constellation and is best found by reference to that star. Through binoculars, M27 appears as a misty patch, but most telescopes will reveal its characteristic dumbbell shape.

Close to the border with Sagitta, and west of the arrow's "flight," is 4, which lies in a curve of faint stars that form part of a loose open cluster. In binoculars, this curve and a straight line of six stars nearby have the appearance of a coat hanger, and the cluster is often referred to as the "Coathanger."

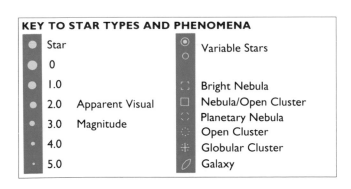

KEY TO STAR TYPES AND PHENOMENA

●	Star	◉	Variable Stars
●	0	◎	
●	1.0		
●	2.0	Apparent Visual	⬓	Bright Nebula
●	3.0	Magnitude	☐	Nebula/Open Cluster
·	4.0		⌣	Planetary Nebula
·	5.0		⁘	Open Cluster
			⁘	Globular Cluster
			⬭	Galaxy

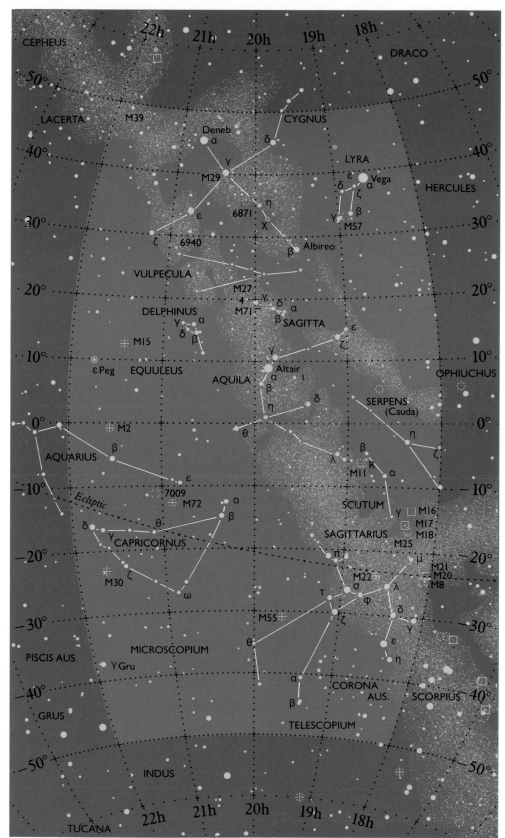

September Stars

As September progresses Pegasus, the Winged Horse, continues chasing the stars of the Summer Triangle toward the western horizon. This signals the approach of fall in the Northern Hemisphere, and of spring in the Southern.

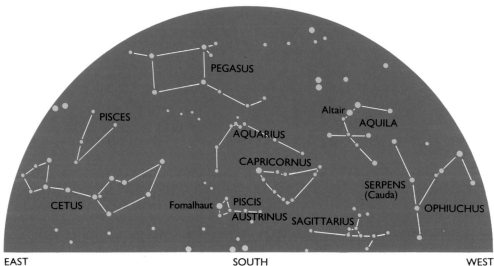

EAST   SOUTH   WEST

▲ **Northern Hemisphere View of the night sky in mid-latitudes looking south at about 11 p.m. local time on about September 7.**
This sky view reveals Altair fairly high in the west and Fomalhaut low in the south. Much of the sky is occupied by rather barren constellations.

▼ **Southern Hemisphere View of the night sky in Australia and southern Africa looking north at about 11 p.m. local time on about September 7.**
Vega and Deneb are now close to the horizon. Pegasus is the most prominent of the rising constellations.

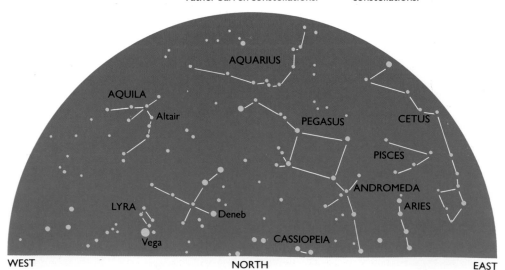

WEST   NORTH   EAST

Aquarius, the Water-Bearer

This is a sprawling zodiacal constellation, but a faint one, with leading stars of only about the third magnitude. It adjoins other "watery" constellations, Pisces, Cetus, Piscis Austrinus, and Capricornus (Fishes, Whale, Southern Fish, and Sea Goat).

The brightest star, Alpha, can perhaps best be found by extending a diagonal of the Square of Pegasus from Alpha Andromedae through Alpha Pegasi. Once Alpha is found, the adjacent stars Gamma, Zeta, and Eta are easy to spot. The pattern of these four stars represents the mouth of the water jar.

Beta is found just off the extended diagonal through Pegasus. Due north of Beta, and forming a right-angled triangle with Alpha and Beta, is M2. This globular cluster is one of the brightest in the heavens, visible as a misty patch in binoculars and resolved into stars in larger telescopes.

Also worth searching for in Aquarius are two fine planetary nebulae. The brighter of the two is NGC7009, which lies south of a line through Beta and Epsilon, forming a right-angled triangle with them. Seen through larger telescopes, NGC7009 appears to have a ring around it rather like the planet Saturn, and it is called the Saturn Nebula. In the same binocular field of view as the Saturn Nebula is M72, another fairly bright globular cluster.

The other planetary nebula, NGC7293, is called the Helix Nebula because of its spiral appearance in long-exposure photographs. In small telescopes, it looks like a faint smoke ring.

Capricornus, the Sea Goat *See page 58.*

Cygnus, the Swan *See page 84.*

Delphinus, the Dolphin *See page 58.*

Grus, the Crane

This is one of the "southern birds," with outstretched wings and a long neck. The long, slightly curving line of stars that marks the neck and body line are readily spotted looking south from the bright Fomalhaut, in the neighboring Piscis Austrinus.

The leading star Alpha has a magnitude of 1.7, slightly brighter than the second-magnitude Beta. These two stars provide a nice color contrast, Alpha being bluish white, Beta being distinctly orange.

Lambda and Iota also have an orange tinge. Delta and Mu are naked-eye doubles, but their pairs of stars are, in reality, far apart.

The area between Theta and the border with the neighboring constellation Phoenix is well worth exploring in telescopes because it contains a veritable cluster of galaxies.

Lacerta, the Lizard

This northern constellation is small and faint, with only its leading star Alpha above the fourth magnitude. Its zigzag of stars, however, overlaps a section of the Milky Way between Cassiopeia and Cygnus, and so it contains glittering starscapes that make fine sweeping with binoculars.

One or two fairly bright open clusters are visible in small telescopes, including NGC7243. This one forms an equilateral triangle with Alpha and Beta.

Of great astronomical interest in the constellation, but far too faint to be seen in ordinary amateur instruments, is the object BL Lacertae. It is a variable galaxy, which changes noticeably in brightness in a relatively short time. Astronomers now think it is a quasar.

Pegasus, the Flying Horse *See page 62.*

Phoenix, the Phoenix *See page 62.*

Pisces, the Fishes *See page 63.*

Piscis Austrinus, the Southern Fish

This small constellation is dominated by a single bright star, the first-magnitude Fomalhaut (Fish's Mouth). No other stars are above the fourth magnitude. Although pure white Fomalhaut rates as only the eighteenth brightest star in the heavens, it is easy to spot because it occurs in a region devoid of any other really bright stars.

Perhaps the most interesting star in this otherwise undistinguished constellation is one called Lacaille 9352. Of the eighth magnitude, it is found south of Pi. At a distance of some 11.7 light-years, it lies relatively close to Earth and shows a substantial proper motion. It moves noticeably across the background of the other stars over a relatively short period.

Vulpecula, the Fox *See page 59.*

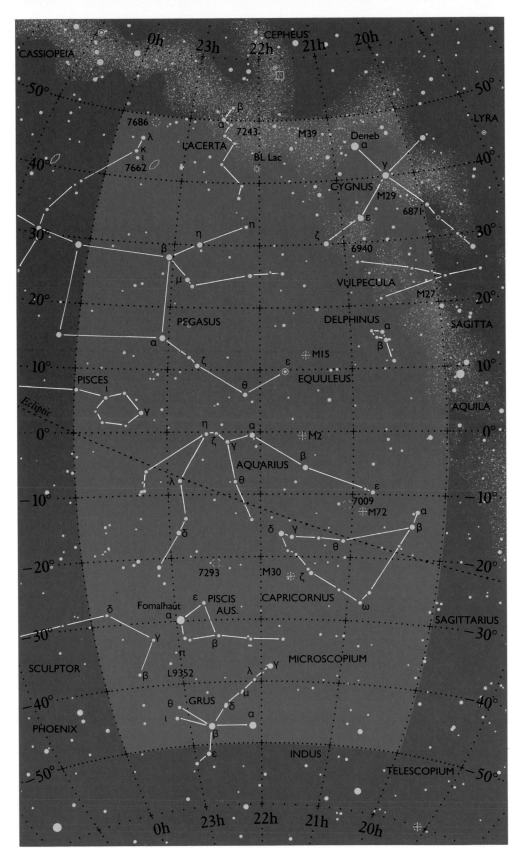

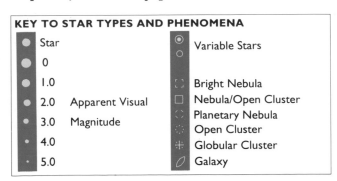

KEY TO STAR TYPES AND PHENOMENA

●	Star	◉	Variable Stars
●	0	○	
●	1.0		
●	2.0 Apparent Visual	⬚	Bright Nebula
●	3.0 Magnitude	▢	Nebula/Open Cluster
●	4.0	◇	Planetary Nebula
·	5.0	⁛	Open Cluster
		✳	Globular Cluster
		⬿	Galaxy

OCTOBER STARS

The October skies are dominated by the unmistakable Pegasus. With few bright stars, the other constellations are harder to trace. This month is good for observing our galactic neighbor, the Great Spiral in Andromeda.

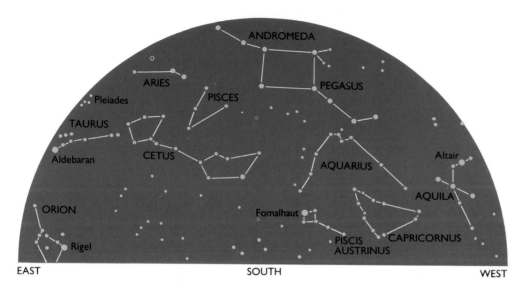

EAST   **SOUTH**   **WEST**

▲ **Northern Hemisphere View of the night sky in mid-latitudes looking south at about 11 p.m. local time on about October 7.**
The Square of Pegasus occupies center stage, while Orion is just rising in the east. Aldebaran and Altair are at a similar altitude.

▼ **Southern Hemisphere View of the night sky in Australia and southern Africa looking north at about 11 p.m. local time on about October 7.**
Pegasus sits in the mid-sky, and Orion is appearing over the eastern horizon. Fomalhaut is prominent near the zenith.

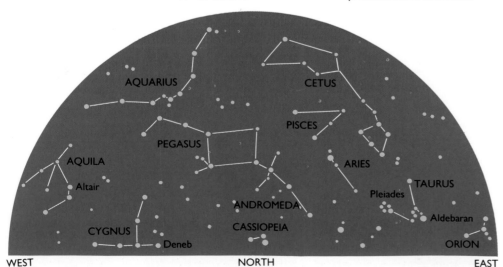

WEST   **NORTH**   **EAST**

Andromeda *See page 78.*

Aquarius, the Water-Bearer *See page 60.*

Aries, the Ram *See page 64.*

Cassiopeia *See page 80.*

Cetus, the Whale *See page 64.*

Eridanus *See page 66.*

Grus, the Crane *See page 60.*

Lacerta, the Lizard *See page 61.*

Pegasus, the Flying Horse

This is a large and distinctive constellation, which dominates fall skies in the Northern Hemisphere. Its four leading stars form one of the most recognizable patterns in the heavens – the Great Square of Pegasus.

Once located, the Square serves as a useful signpost to other constellations. To the naked eye, the area enclosed by the Square is remarkably empty of stars. You would be hard-pressed to see more than a dozen.

Pegasus shares one of the stars in the Square with the neighboring constellation Andromeda – Alpha Andromedae. (Its Arab name is Alpheratz.) Going clockwise around the Square, the three other stars are Beta (Scheat), Alpha (Markab), and Gamma (Algenib).

All of these stars are of the second magnitude, with Alpha Andromedae being the brightest. Beta is a variable star, although its variation in brightness is not easy to follow. It is a red giant star, and its reddish tinge, apparent even to the naked eye, contrasts with the pure white of Alpha.

The noticeably yellow Epsilon (Enif) lies some distance from the Square, roughly in line with Gamma and Alpha. It is fractionally brighter than Alpha and is a double, whose fainter component can be spotted in binoculars.

Among deep-sky objects in Pegasus, the globular cluster M15 is outstanding. It lies in line with Theta and Epsilon, a fraction too faint to be seen with the naked eye, but readily visible in binoculars and small telescopes. Larger instruments will resolve the outer regions into individual stars. (See the map on page 65 for location.)

Phoenix, the Phoenix

The second-magnitude Alpha marks the eye of this "southern bird," which in mythology was reincarnated from its own ashes. It is a fairly large, but rather barren constellation. Beta is a double; both components are yellow and equally bright. Zeta is another double; the brightness of the brighter component varies slightly, and instruments show that it is a close eclipsing binary.

Pisces, the Fishes

This is a large but rather uninteresting constellation of the zodiac, formed of a string of faint stars little brighter than about the fourth magnitude.

One of the two fishes the constellation depicts is marked by a circle of stars south of the Great Square of Pegasus. This is the easiest part of the constellation to identify. The curved line of stars from there to the leading star Alpha is not so easy to follow; nor is the line from Alpha to Eta and beyond.

Alpha is a close double star, separated in larger telescopes into bluish-white components. Zeta is a much wider double, with twin pale yellow stars.

Among several galaxies in Pisces, M74 is the brightest. Located close to Eta, it can just be picked up by small telescopes.

Piscis Austrinus, the Southern Fish See page 61.

Sculptor, the Sculptor

With Phoenix and Fornax, this constellation appears in a relatively barren region of the mid-southern skies. It is best found by reference to the bright Fomalhaut in the neighboring Piscis Austrinus.

Sculptor is notable mainly for its galaxies, several of which (if you can find them!) can be glimpsed in binoculars. Larger telescopes will pick up many more. Definitely in binocular range is NGC253, visible as an elongated misty patch about a third of the way along a line from Alpha toward Beta Ceti to the north. Larger telescopes reveal NGC253 to be a spiral galaxy, much like our own but smaller.

NGC253 is the brightest member of a cluster of galaxies in the region, known as the Sculptor group. They lie about 10 million light-years away.

Triangulum, the Triangle See page 65.

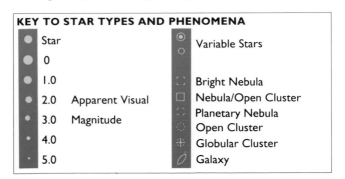

KEY TO STAR TYPES AND PHENOMENA

●	Star	◉	Variable Stars
●	0	○	
●	1.0		
●	2.0	Apparent Visual	⬚	Bright Nebula
●	3.0	Magnitude	▫	Nebula/Open Cluster
•	4.0		⬭	Planetary Nebula
·	5.0		⬚	Open Cluster
		✠	Globular Cluster
		⬭	Galaxy

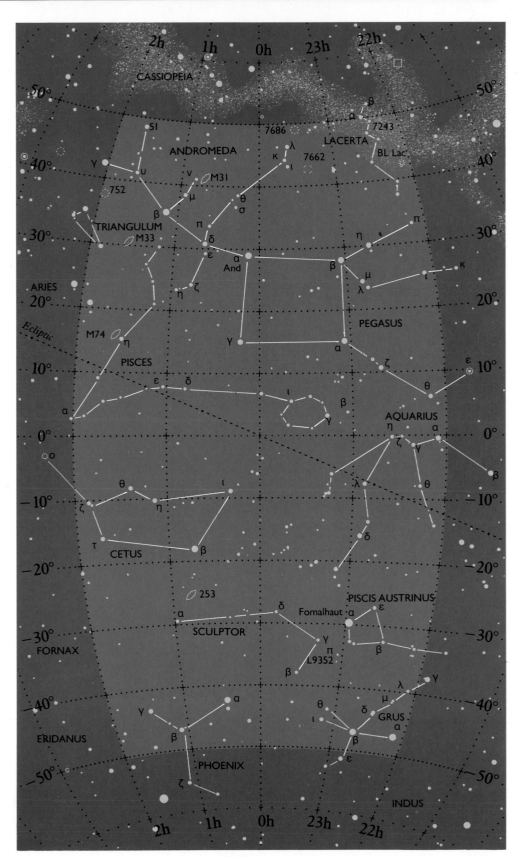

November Stars

As Pegasus begins its slow descent in the west, Taurus and Orion begin climbing to prominence in the east. Orange Aldebaran, the Hyades, and the Pleiades clusters are becoming well placed for observation in both hemispheres.

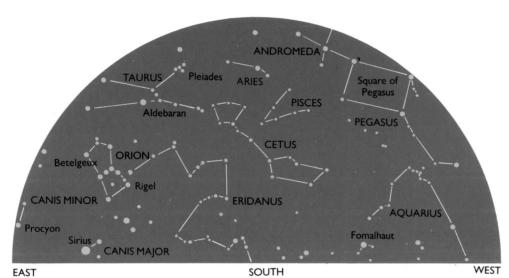

EAST   SOUTH   WEST

▲ **Northern Hemisphere View of the night sky in mid-latitudes looking south at about 11 p.m. local time on about November 7.** Faint constellations such as Eridanus and Cetus cover much of the sky. The interest lies in the east, in sparkling Orion and Taurus.

▼ **Southern Hemisphere View of the night sky in Australia and southern Africa looking north at about 11 p.m. local time on about November 7.** Pegasus and Andromeda are about to set in the west. In the east, Aldebaran, Betelgeux, and Rigel form a nice triangle.

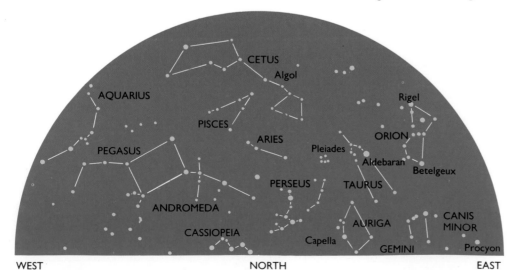

WEST   NORTH   EAST

Andromeda *See page 78.*

Aries, the Ram

This small constellation of the zodiac has just two noticeably bright stars, the second-magnitude Alpha and the somewhat less bright Beta. This pair can probably best be found by reference to the Great Square of Pegasus. Alpha is also at the same declination as the Pleiades cluster in the neighboring Taurus.

Gamma, due south of Beta, is a fine double, visible in small telescopes as a pair of equally bright bluish-white stars.

Cetus, the Whale

Although it is one of the largest constellations in the heavens, Cetus is not easy to trace because of its lack of bright stars. Only Alpha in the Whale's head and Beta in its tail are above the third magnitude.

However, both these and the other main stars lie more or less on a line between Antares in Taurus and Fomalhaut in Piscis Austrinus.

In the Whale's head, Alpha is a wide binocular double. It is a red giant, whose orange tinge is discernible to the naked eye and very obvious through binoculars. A telescope is needed to split the adjacent bright double star, Gamma, into its two components, which provide a marked color contrast of bluish-white and yellow.

In the tail region of Cetus, Tau is interesting in that it is a star of a similar type to the Sun. Some 12 light-years away, it is one of the closest stars that astronomers think could have a planetary system like the Sun's. It is a favorite target for SETI (searches for extraterrestrial intelligence) projects.

Historically, the most important star in the constellation is Omicron, also called Mira, meaning "wonderful." It is the prototype long-period variable star, a huge red giant that varies widely in brightness over a period of several months or a year or more.

Mira Ceti varies between magnitudes of about 3 and 10 every 11 months or so. It is found midway between Gamma and the distinctive pair Zeta and Chi, and is easily seen when it is near maximum, but difficult to find when it is near minimum.

A number of galaxies lie near Delta, the brightest of which, M77, can be sighted in binoculars as a hazy patch. A telescope will show it is a face-on spiral with bright, widely spaced arms. It is one of the most remote Messier objects, lying at a distance of over 50 million light-years.

Eridanus *See page 66.*

Fornax, the Furnace *See page 66.*

Pegasus, the Flying Horse *See page 62.*

Perseus *See page 67.*

Phoenix, the Phoenix *See page 62.*

Pisces, the Fishes *See page 63.*

Sculptor, the Sculptor *See page 63.*

Triangulum, the Triangle

With its three main stars arranged in a neat triangle, this constellation is well named. However, it is far from bright, with just two stars, Alpha and Beta, of about the third magnitude. The third main star, Gamma, makes up the third corner of the triangle, along with two close fainter companions.

Triangulum's main claim to fame is the beautiful face-on spiral galaxy M33, which lies about halfway between Alpha Trianguli, and Beta Andromedae in the neighboring constellation. It is just about visible to the naked eye on a really dark night and can readily be seen as a hazy patch in binoculars. Larger telescopes, however, are needed to bring out its wide-open spiral arms.

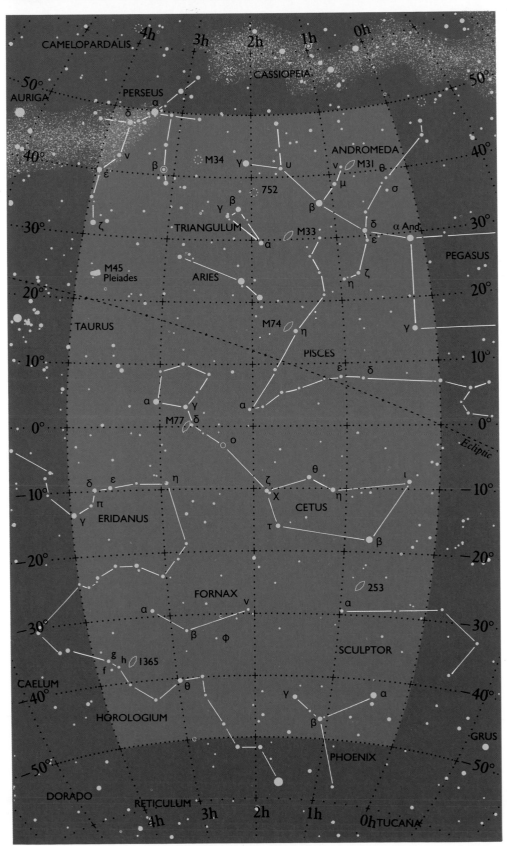

KEY TO STAR TYPES AND PHENOMENA

Star		⊙	Variable Stars
0		○	
1.0		▢	Bright Nebula
2.0	Apparent Visual	▢	Nebula/Open Cluster
3.0	Magnitude	◇	Planetary Nebula
4.0		⊙	Open Cluster
5.0		✳	Globular Cluster
		⬭	Galaxy

DECEMBER STARS

A hexagonal arrangement of bright stars – Sirius, Rigel, Aldebaran, Capella, Pollux, and Procyon – in Canis Major, Orion, Auriga, Gemini, and Canis Minor make spectacular stargazing in this month and the next.

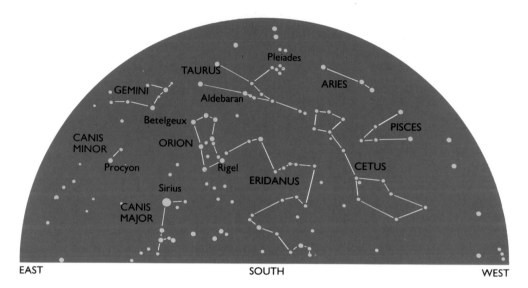

EAST   SOUTH   WEST

▲ **Northern Hemisphere View of the night sky in mid-latitudes looking south at about 11 p.m. local time on about December 7.**
The meridian neatly splits the sky into two. The eastern half sparkles with bright stars and provides a marked contrast to the lackluster western half.

▼ **Southern Hemisphere View of the night sky in Australia and southern Africa looking north at about 11 p.m. local time on about December 7.**
The December hexagon of bright stars is stunning in the east. The Pleiades cluster is in an ideal viewing position.

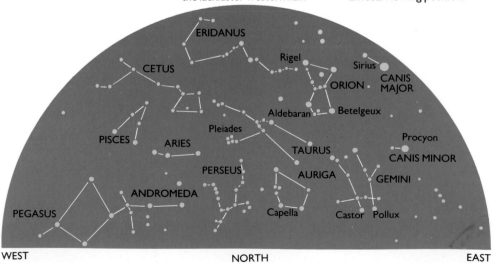

WEST   NORTH   EAST

Aries, the Ram *See page 64.*

Auriga, the Charioteer *See page 44.*

Carina, the Keel *See page 42.*

Cetus, the Whale *See page 64.*

Columba, the Dove *See page 45.*

Eridanus

One of the largest of the constellations, Eridanus represents a river. The ancient Middle Eastern astronomers associated it with their life-giving rivers, the Euphrates and the Nile.

With broad meanders, Eridanus winds itself through the heavens. It "rises" at Beta, near the bright Rigel in Orion, and works its way lazily toward the brilliant Alpha far south. The Arab name for Alpha, Achernar, means "end of the river." Achernar, of magnitude 0.5, is the ninth brightest star in the sky.

Heading "upstream" from Achernar, Theta is perhaps the first interesting star. It is a fine double, which small telescopes will separate as a bluish-white pair.

The other interesting stars lie in the northern reaches of the river. Omicron² is a wide and easy double. The fainter of its components is a white dwarf star and is the easiest of its type to see. Larger telescopes reveal that the white dwarf itself has a companion, a red dwarf.

Delta and Pi are both distinctly orange and form an obvious triangle with Epsilon. Epsilon lies relatively close to Earth (10.7 light-years) and is a similar type of star to the Sun. Astronomers have long thought that it might have a planetary system like Earth's.

Fornax, the Furnace

This is one of the faint constellations in mid-southern skies, which, like the neighboring Sculptor, lacks prominent stars. Only Alpha is brighter, just, than the fourth magnitude. Mostly, Fornax is encircled by one of the broad meanders of Eridanus.

Although binoculars find little of interest, small telescopes begin to reveal the riches Fornax possesses. Zeroing in on the faint twin stars Lambda reveals a dense star cluster nearby. Larger instruments show it to be an irregular galaxy, called the Fornax System. It is a dwarf galaxy, like the Magellanic Clouds.

On the edge of the constellation, close to the triangle of stars, g, f and h Eridani, telescopes reveal the Fornax Cluster of galaxies. Larger instruments show the Cluster to have about 18 members, one of which is NGC1365, one of the brightest barred-spiral galaxies in the heavens.

At a distance of about 55 million light-years, the Fornax Cluster is one of the nearest clusters of galaxies.

Lepus, the Hare *See page 45.*

Orion *See page 86.*

Perseus

This bright northern constellation has some splendid stars and glorious star fields, since it straddles the Milky Way. The brightest star, Alpha (also called by its Arab name Mirfak), is surrounded by a myriad of faint stars, and the group prettily fills the whole field in binoculars.

Binoculars will also reveal the rare beauty of one of the showpiece features of Perseus, the famous Double Cluster. Also called the Sword Handle, it is easily located midway between Gamma Persei and Delta Cassiopeiae in the neighboring constellation. It is a pair of glorious open clusters, each of which would be beautiful by itself. The keen-eyed should see them without optical aid on a dark night. On star charts they are usually separately identified as h and Chi Persei, or NGC 869 and 884.

The other showpiece feature in Perseus is Beta. Its Arab name Algol, or the Winking Demon, describes it well. For most of the time, it shines steadily at about magnitude 2, but about every 2½ days it fades over a few hours to about magnitude 3.5 before regaining its former brightness. Algol is a classic example of an eclipsing binary and was the first of its type to be recognized.

In the same binocular field as Algol is another open cluster, M34, which is just visible to the naked eye.

Taurus, the Bull *See page 92.*

Triangulum, the Triangle *See page 65.*

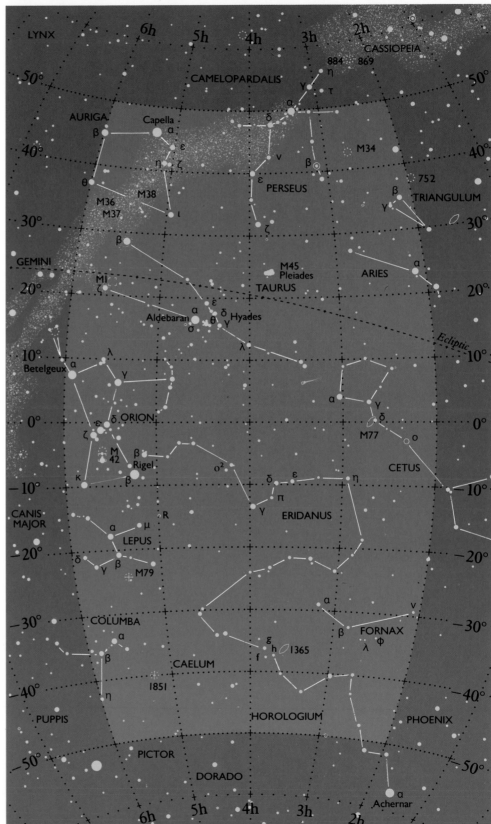

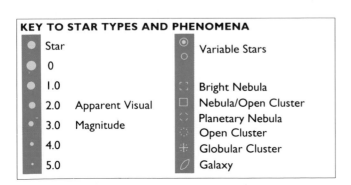

KEY TO STAR TYPES AND PHENOMENA

Star		Variable Stars
0		
1.0		Bright Nebula
2.0	Apparent Visual	Nebula/Open Cluster
3.0	Magnitude	Planetary Nebula
		Open Cluster
4.0		Globular Cluster
5.0		Galaxy

▲ The brilliant star Capella in Auriga, which is the third brightest star in northern skies, after Arcturus in Boötes and Vega in Lyra, and the sixth brightest star in the whole heavens.

STAR ANATOMY

Stars are great globes of searing hot gas, which generate enormous energy in their interior and radiate it into space. All the stars except one lie very far away from us, at distances measured in million-millions of miles. The nearby star is the one we call the Sun.

The Sun is very special to us, but in the Universe as a whole it is very ordinary. It is not particularly big nor particularly bright. There are many millions of other stars in the night sky like it. But we know of millions of other stars that are quite unlike it: red giant and supergiant stars of truly gigantic dimensions; stars tinier than planets; stars many thousands of times brighter than the Sun; and stars that fluctuate wildly in brightness.

Magnitudes of brightness

As a quick glance at the night sky shows, the stars vary widely in brightness. Stars like Sirius, Canopus, Alpha Centauri, and Arcturus shine like celestial beacons, while others are scarcely visible to the naked eye.

In about 150 B.C., the Greek astronomer Hipparchus devised a scale of "magnitudes" for grading the brightness of stars. He allotted a magnitude of 1 to the brightest stars, and a magnitude of 6 to those just visible to the naked eye.

This system forms the basis of the magnitude scale astronomers still use today. But they have refined and extended it. On the scale, a star of the first magnitude (magnitude 1) is 100 times brighter than a star of the sixth magnitude. This means that a star of the first magnitude is about two-and-a-half times brighter than one of the second magnitude, which is two-and-a-half times brighter than one of the third magnitude, and so on.

Because brightness can now be accurately measured with instruments such as photometers, magnitudes can be expressed accurately to one or two decimal places. Deneb, for example, has a magnitude of 1.25. (In this book, however, magnitudes are usually expressed to the first decimal place for simplicity.)

To express the brightness of stars beyond naked-eye range, the magnitude scale is extended forward beyond magnitude 6. The closest star to us after the Sun, Proxima Centauri, has a magnitude of 11.1. The most powerful telescopes are able to detect stars as faint as magnitude 25.

To measure the brightness of exceptionally bright stars such as Sirius, the scale is extended backward from

BRIGHTEST STARS

Brightest Stars	Designation	App. magnitude	Absolute magnitude	Spectral type	Distance (Light-years)
Sun	—	−26.7	4.8	G2	—
Sirius	α Canis Majoris	−1.5	1.4	A1	8.8
Canopus	α Carinae	−0.7	−4.7	F0	196
Rigil Kent	α Centauri	−0.3	4.4	G2	4.3
Arcturus	α Boötis	−0.1	−0.2	K1	37
Vega	α Lyrae	0.0	0.5	A0	26
Capella	α Aurigae	0.1	−0.6	G8	46
Rigel	β Orionis	0.1	−7.0	B8	815
Procyon	α Canis Minoris	0.4	2.7	F5	11
Achernar	α Eridani	0.5	−2.2	B5	127
Hadar	β Centauri	0.6	−5.0	B1	391
Altair	α Aquilae	0.8	2.3	A7	16
Betelgeux	α Orionis	0.8	−6.0	M2	652
Aldebaran	α Tauri	0.8	−0.7	K5	68
Acrux	α Crucis	0.9	−3.5	B1	261
Spica	α Virginis	1.0	−3.4	B1	261
Antares	α Scorpii	1.0	−4.7	M1	424
Pollux	β Geminorum	1.1	1.0	K0	36
Fomalhaut	α Piscis Austrini	1.2	1.9	A3	23
Deneb	α Cygni	1.3	−7.3	A2	1630
Mimosa	β Crucis	1.3	−4.7	B0	490
Regulus	α Leonis	1.4	−0.6	B7	85

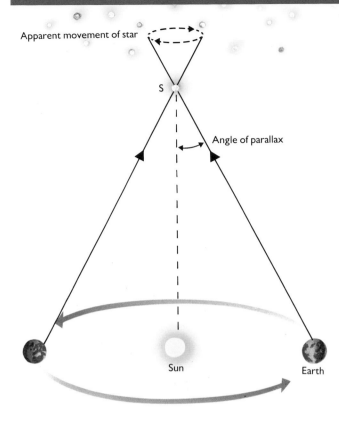

Apparent movement of star

S

Angle of parallax

Sun

Earth

▲ We can measure the distance to a few nearby stars using the principle of parallax. We note the different positions of the star S against the background of distant stars when viewed from opposite ends of the Earth's orbit. We can work out the star's distance using trigonometry.

magnitude 1 into negative values. So Sirius has a magnitude of –1.46. On this same scale, the brightest planet, Venus, can reach a magnitude of –4.4; the Moon is –12.7, and the Sun –26.7.

The brightness of a star that we observe from Earth is only an apparent brightness. It bears little relation to the star's true brightness: a nearby truly dim star will appear brighter than a very distant truly bright star.

To compare the true brightness of stars, we would have to view them from the same distance. And that is the basis of the scale of true or absolute magnitude astronomers use. They define the absolute magnitude of a star as the magnitude they would observe if the star were at a distance of 10 parsecs (see below). On this absolute scale, Sirius now rates only 1.4, whereas Antares rates –4.7. Some stars, such as Rigel, rate more than –7.

Distance to the stars

No matter how powerful a telescope you have, you can never see the stars other than as pinpoints of light. That is because they are so far away that their light takes years to reach us. Indeed, we often express the distance to a star in terms of the number of years it takes for its light to reach us. The light from Sirius takes 8.8 years to reach us, so we can say that Sirius is 8.8 "light-years" away, 1 light-year being the distance light travels in a year (5.9 million million miles, 9.5 million million kilometers).

In this book we express stellar distances in terms of light-years. But professional astronomers generally use a unit called the parsec, which is equivalent to about 3.3

Star

Actual path of star

Radial motion

Proper motion

Earth

▲ All the stars are moving relative to us. For just a few, we can detect their proper motion, or motion across our line of sight.

NEAREST STARS

Nearest Stars	Distance (Light-years)	Proper motion ("/yr)	App. magnitude	Absolute magnitude	Spectral type
Proxima Centauri	4.3	3.9	11.1	15.5	M5
α Centauri A	4.3	3.7	–0.3	4.4	G2
α Centauri B	4.3	3.7	1.3	5.7	K5
Barnard's Star	5.9	10.3	9.5	13.3	M5
Wolf 359	7.6	4.7	13.5	16.7	M8
Lalande 21185	8.1	4.8	7.5	10.5	M2
Sirius A	8.8	1.3	–1.5	1.4	A1
Sirius B	8.8	1.3	8.7	11.6	A
Luyten 726–8A	8.9	3.4	12.5	15.3	M5
UV Ceti	8.9	3.4	13.0	15.8	M6
Ross 154	9.4	0.7	10.6	13.3	M4
Ross 248	10.3	1.6	12.3	14.8	M6
ε Eridani	10.8	1.0	3.7	6.1	K2
Luyten 789–6	10.8	3.3	12.2	14.6	M7
Ross 128	10.8	1.4	11.1	13.5	M5
61 Cygni A	11.1	5.2	5.2	7.6	K5
61 Cygni B	11.1	5.2	6.0	8.4	K7
ε Indi	11.2	4.7	4.7	7.0	K5
Procyon A	11.4	1.3	0.4	2.7	F5
Procyon B	11.4	1.3	10.7	13.0	F
Struve 2398 A	11.5	2.3	8.9	11.2	M4
Struve 2398 B	11.5	2.3	9.7	12.0	M5

▼ The size of stars varies enormously. Our star the Sun is classed as a yellow dwarf. It is small compared with hotter blue-white stars like Spica. But, some 5,000 million years hence, it will swell up into a giant. Bigger stars expand into supergiants, with diameters measured in many hundreds of millions of miles.

Yellow dwarf

Blue-white star

Red giant

Supergiant

Larger giant

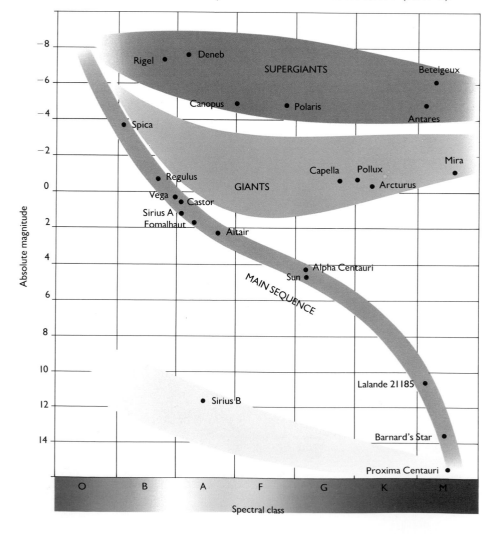

▲ A star's spectrum provides a means of classifying it. The top spectrum is typical of white stars like Deneb, with a surface temperature of about 20,000°F (11,000°C). The bottom spectrum is typical of red stars like Betelgeux, with a surface temperature of only about 5,000°F (3,000°C).

◄ The Hertzsprung-Russell diagram relates the absolute brightness and the spectral class of stars. Most stars lie on a diagonal, called the Main Sequence.

light-years. (1 parsec is the distance at which a star would show an angle of parallax of one second of arc, or 1/3600 of a degree; see diagram, left.)

The message in starlight

Although the stars lie very far away, we know a lot about them. We can often tell how hot they are, what they are made of, how fast they are moving, and so on. All this information – and more – we can extract from the feeble light that reaches us through space. We get at it by passing the light through the most important astronomical instrument after the telescope – the spectroscope (or spectrograph). This instrument splits up starlight into a spectrum, a spread of different light wavelengths, or colors, from violet-blue (shortest waves) to red. A series of dark lines crosses the spectrum at intervals.

Stars have different kinds of spectra. For example, some are brightest in blue, some in yellow, some in red, and so on. Their color intensity is a measure of their surface temperature. Spectra also differ in the number and arrangement of dark lines. From the position of the lines we can tell, for example, the star's composition. We can also tell how fast and in which direction the star is moving. If the star is traveling toward us, the spectral lines are shifted bodily toward the blue end of the spectrum. If it is traveling away from us, the spectral lines are shifted toward the red end. The amount of shift is a measure of the star's speed.

The spectrum provides a useful way of classifying a star, because stars of a similar type have a similar kind of

▲ This dense star field is in Cygnus. The brightest star in the picture is the first-magnitude Deneb. This star has an absolute magnitude of at least −7.3 and is one of the most luminous stars known.

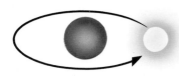 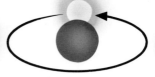 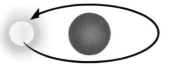 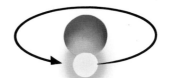 

Brightness   Primary eclipse   Secondary eclipse

Eclipsing binaries

The star Beta Persei is named Algol, or "The Winking Demon," because every 69 hours it suddenly dims from its regular magnitude of about 2 to about magnitude 4. The first person to explain Algol's odd behavior was the deaf-mute English astronomer John Goodricke in 1783.

Algol behaves as it does because it belongs to a class of stars called eclipsing binaries. It consists of two stars orbiting around each other: one large and dim, and the other small and bright. The plane of their orbit is in our line of sight, so periodically they pass in front of, or eclipse, each other. When the dim one eclipses the bright one, the brightness suddenly drops markedly. But there is only a slight drop in brightness when the dim one is eclipsed.

---

spectrum. There are ten main kinds of spectra, or spectral classes, designated O, B, A, F, G, K, M, R, N, and S, in order of decreasing surface temperatures, which run from about 70,000° to below 5,000°F (40,000°–3,000°C). (This sequence of spectral classes can be remembered by the never-to-be-forgotten mnemonic "Oh Be A Fine Girl Kiss Me Right Now Sweetie"!)

The spectral class and absolute magnitude (see page 69) are two of the basic characteristics of a star. When they are plotted against one another, the stars fall into distinct groups. The resulting plot is called the Hertzsprung-Russell (H-R) Diagram, after the astronom___ who first devised it, Ejnar Hertzsprung in Denmar___ Henry Russell in the United States.

A version of the H-R Diagram for some of the ___ known stars appears on the opposite page. Most sta___ including our own Sun, lie on the diagonal band calle___ the Main Sequence. Stars spend most of their life, shining steadily, on the Main Sequence.

The variable stars
Most of the stars we see in the sky shine steadily. (They "twinkle," it is true, but this is because of air currents in the Earth's atmosphere.) However, some stars, called variable stars, vary noticeably in brightness over a short or long period.

The star Beta Persei, or Algol, varies in brightness every 69 hours. But that is because it is a two-star, or binary, system in which one star periodically eclipses another. We know of many "eclipsing binaries" like it.

Other stars vary in brightness because of processes going on inside them. The first to be discovered was Omicron Ceti, or Mira (meaning "Wonderful"), which varies between magnitudes 3 and 10 in about 330 days. Mira typifies what are known as long-period variables, which are believed to be pulsating red giant stars.

Other variable stars vary over much shorter periods, for example, Delta Cephei, which is the prototype star for the Cepheid variables (see below).

Novae
Sometimes a bright new star suddenly appears in the sky, and we call it a nova (meaning "new"). In fact, it is not a new star at all, but an existing dim star that has suddenly in___ ___amatically in brightness, typically by m___ ___imes. This can happen over only ___quent decline in brightness to

"new star" occurrences are ___ when supergiant stars blast ___h throes (see page 72).

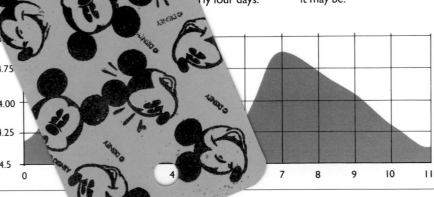

___any other stars vary ___rly in the same way, and ___e known as the ___ds. The typical, or ___Cepheids have periods ___m one to about 50 ___ is a Cepheid with a ___rly four days.

In 1912 Henrietta Leavitt in the USA discovered that the period of a Cepheid is directly related to its absolute magnitude, or luminosity. This law enables us to measure the distance to a Cepheid wherever it may be.

Star Life and Death

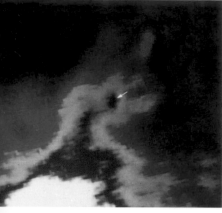

The lifetime of a star is measured in hundreds or thousands of millions of years, so we cannot trace the life cycle of any particular star from beginning to end. However, we can study stars of different ages and so piece together the life cycle of a typical star.

▲ A newborn star (arrowed) in the nebula Barnard 5 in Perseus, spotted by the infrared astronomy satellite IRAS. At present, the temperature and pressure in this protostar's interior are not yet high enough to trigger off nuclear reactions.

Stars are born in nebulae – the billowing clouds of dust and gas (mainly hydrogen) that exist between the stars in the galaxies. Star formation begins when a particularly dense part of a nebula begins to condense, or starts to shrink, under the influence of gravity.

No one knows for sure what triggers the process. It could be the shock waves that ripple through space after another star explodes and dies. That would be neat: the death of one star triggering the birth of another.

As the nebula shrinks under gravity, it releases energy, which appears as heat. The smaller the collapsing nebula (or "protostar") becomes, the hotter it gets. Event-

◄ The Dumbbell Nebula (M27) in Vulpecula. The distinctive shape of this planetary nebula is caused by a shell of expanding gas and dust, puffed out by the central star millions of years ago.

► The remnants of a supernova witnessed by the celebrated Danish astronomer Tycho Brahe in 1572. It is an X-ray image, taken by the German Rosat (Roentgen satellite). Tycho is one of the few supernova remnants whose age we know precisely from recorded observations.

ually, temperatures inside reach over 18,000,000°F (10,000,000°C). Then the nuclei (centers) of the hydrogen atoms begin to fuse, or join together. This nuclear fusion process produces fantastic amounts of energy as light, heat, and other kinds of radiation. The radiation eventually makes its way to the surface and escapes into space, and the body starts to shine, as a star. What happens to the star from then on depends on its mass.

A star like the Sun shines steadily for about 10,000 million years. Then with its hydrogen fuel exhausted, it swells up into a red giant star with a diameter tens of times bigger than before. In time, the red giant begins to shrink, occasionally puffing off gas which astronomers detect as a planetary nebula. As the dying star shrinks, it heats up, becoming a small, hot white dwarf.

Stars more massive than the Sun meet a more spectacular end. They expand beyond the red-giant stage into a supergiant, with hundreds of times their original diameter. A supergiant is unstable and soon collapses. This triggers off a cataclysmic explosion – a supernova – in which most of the star is blasted into space.

The outcome of a supernova explosion again depends on the original mass of the star. Stars with a mass up to about seven times that of the Sun leave behind a very dense neutron star, typically only about 12–20 miles (20–30 kilometers) across. Stars with even greater mass are virtually crushed out of existence. They leave behind a region of space with such enormous gravity that nothing, not even light, can escape from it. We call such a region a black hole.

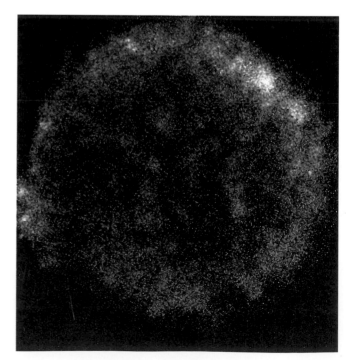

Supernova 1987A

In the early morning of February 20, 1987, astronomer Ian Shelton became the first person for centuries to witness a supernova with the naked eye. He was observing at the Las Campanas Observatory in Chile, South America.

The supernova, SN 1987A, took place in our galactic neighbor, the Large Magellanic Cloud. It occurred when the supergiant star Sanduleak — 69° 202 exploded. It has been the most widely studied of any supernova.

▲ SN 1987A at maximum brightness, outshining every star in the Large Magellanic Cloud. It occurred close to what is ordinarily the highlight of the galaxy, the Tarantula Nebula.

► The expanding ring of material blasted into space during the SN 1987A explosion is shown in the middle of this picture returned by the Hubble Space Telescope. A core of debris at the center is what remains of the exploding star.

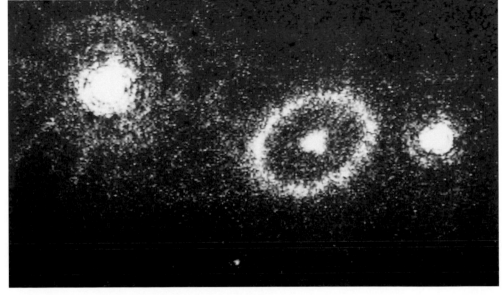

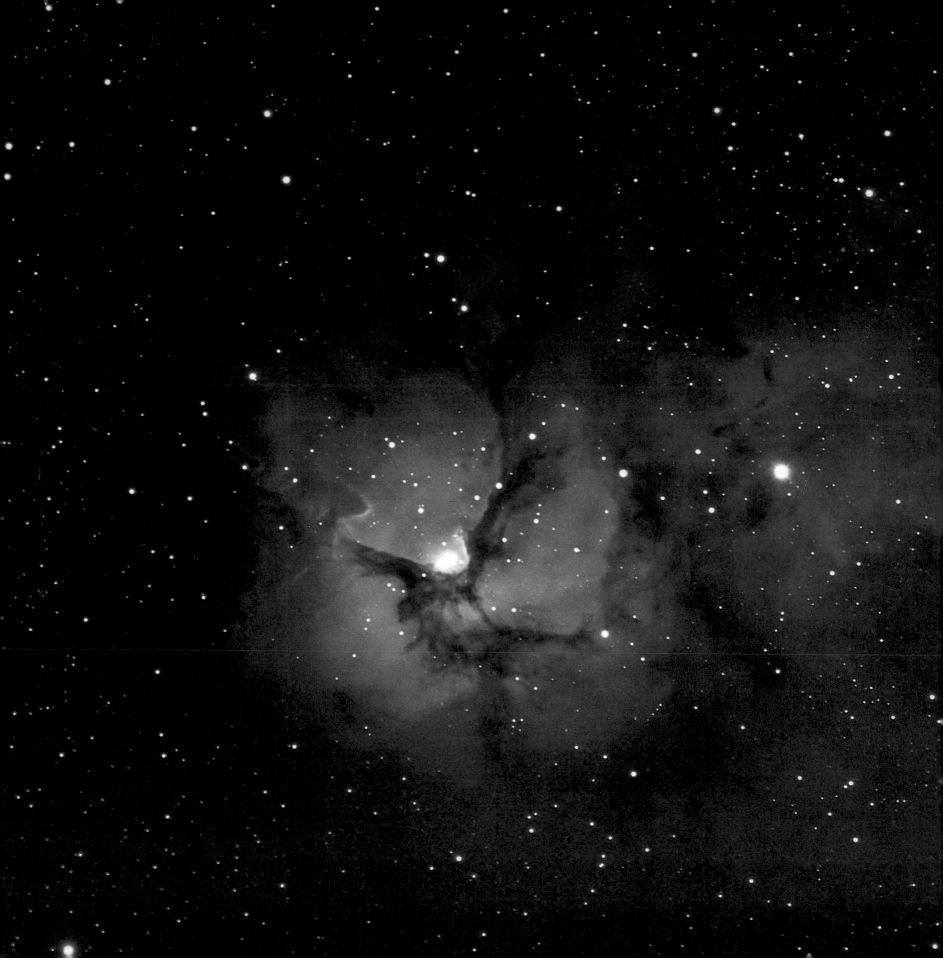

◄ The Trifid Nebula (M20) is one of the many astronomical delights in Sagittarius. A most distinctive nebula, it is named after the dark dust lanes that split it into fan-shaped segments. The Trifid is an unusual combination of an emission nebula (the main red cloud) and a reflection nebula (the fainter blue one).

3

KEY CONSTELLATIONS

Of the 88 constellations that cover the celestial sphere, a number are truly outstanding for one reason or another. In the Southern Hemisphere they include Sagittarius. This constellation is magnificent not so much for its stars, but for the abundance of clusters and nebulae (like the Trifid, left) it contains, and for its dense star clouds. In the Northern Hemisphere, Taurus boasts two exceptional star clusters. Orion, which spans the celestial equator, displays fine stars, brilliant nebulae, and acts as a signpost for locating stars in other constellations.

Sagittarius, Taurus, and Orion are three of the eleven key constellations that we feature in this chapter for more detailed investigation. The others are Andromeda, Cassiopeia, Centaurus and Crux, Cygnus, Scorpius, Ursa Major, and Virgo.

Each constellation (except for Centaurus and Crux) rates a double-page spread, with the constellation map occupying a full page. The descriptive text is prefaced by the fascinating story behind the mythological figure which the constellation purports to resemble.

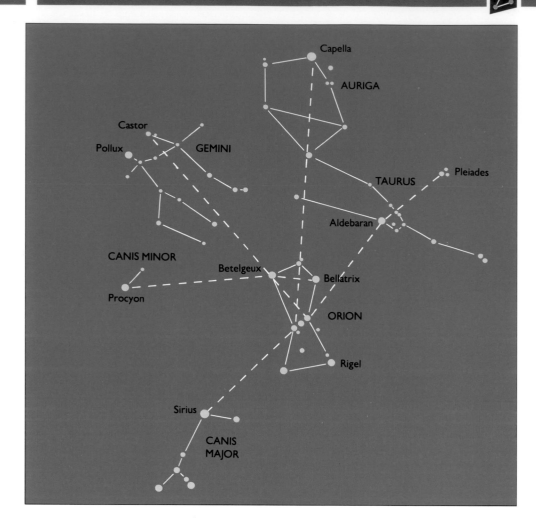

SIGNPOSTS IN THE SKY

The eleven constellations featured in this chapter are among the most spectacular in the heavens. They are drawn from both the northern and the southern hemispheres. Each has more than its fair share of colored stars and variables, doubles and clusters, and nebulae and galaxies. Many of them, moreover, are invaluable in guiding astronomers around the sky.

There is no other celestial signpost in the heavens to rival the Big Dipper, the most prominent part of Ursa Major. Although it boasts no star of the first magnitude, the Big Dipper is quite unmistakable, which is a good start! If you look at the diagram on the left, you can see just how useful it can be.

The four stars that form the cup of the Big Dipper act in pairs as pointers, not only to the Pole Star, Polaris, but also to Deneb in Cygnus, Capella in Auriga, Pollux and twin Castor in Gemini, and Regulus in Leo. Following the curve of the handle brings you to Arcturus, the brightest star in northern skies. And by continuing the curve further, you reach Spica in Virgo.

Orion *(left)* is unmistakable, too. Its shape is distinctive, and its stars are bright, with two – Betelgeux and Rigel – being of the first magnitude. The three stars in Orion's Belt are pointers to brightest-star-in-the-sky Sirius in Canis Major and Aldebaran in Taurus, the fiery eye of the bull. A diagonal through Rigel and Betelgeux

Orion
Orion, too, is an excellent celestial signpost. Spanning the celestial equator, it is a useful

constellation for observers in both hemispheres. The line through the three stars of Orion's Belt south to Sirius

and north to Aldebaran and the Pleiades is particularly striking.

▼ One of the many highlights of the key constellation Orion is the Great Nebula (M42), in which is embedded the

multiple star called the Trapezium. This picture shows clearly the star's four components.

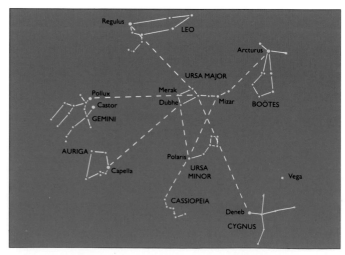

Ursa Major
In the Northern Hemisphere, Ursa Major is indeed a key constellation. The Big Dipper serves as an invaluable signpost for pointing the way to a host of other stars.

leads to Castor and its twin Pollux. Going almost due east from Gamma Orionis, or Bellatrix, leads you to Procyon in Canis Minor. Capella is found on a line running north through the middle of the main star pattern.

The most obvious pointers in far southern skies are Alpha and Beta Centauri, which guide the eye to the Southern Cross, or Crux. Without them, the eye might be tempted to mistake the so-called False Cross, not far away, for the real one. The False Cross is made up of a pair of stars from Vela and a pair from Carina. Unlike the Northern Hemisphere, the Southern Hemisphere has no convenient pole star, nor even pointers in the right direction. The long axis of the Southern Cross, however, points nearly in the right direction.

Epoch 2000

It must be said that the positions of the stars on the celestial sphere on the maps that follow – and on those in Chapter 2 – are not quite accurate. They are shown in the positions they will occupy on January 1, 2000. The year 2000 is the so-called standard epoch of observation. But the difference between the stars' present positions and those for Epoch 2000 is now negligible.

The reason why the positions of the stars change slowly is precession. This is the "wobbling" of the Earth's axis, which causes the celestial poles to describe a circle in space once about every 26,000 years. Because of precession, Deneb in Cygnus will become the pole star in about A.D. 10,000.

▼ This star cloud in the key constellation Sagittarius is the brightest in the sky. Easily seen with the naked eye, it is magnificent in binoculars, like all parts of the Milky Way in Sagittarius.

▲ In the key constellation Andromeda is the incomparable Great Spiral galaxy, on the edge of which is this elliptical galaxy, M32. This is one of the two companions of the Great Spiral.

▼ In the Southern Hemisphere, Centaurus and Crux are key constellations. This photograph shows (*at right*) Alpha and Beta Centauri pointing to the unmistakable Southern Cross.

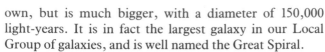

ANDROMEDA

Andromeda was the beautiful daughter of King Cepheus and Queen Cassiopeia. By boasting about her own beauty (see page 80), the Queen upset Poseidon, the sea god, who sent a monster, Cetus, to terrorize the shores of Cepheus's kingdom. To placate Poseidon, the King chained Andromeda to rocks on the shore as a sacrifice. Just as Cetus was about to pounce on her, the hero Perseus happened by and killed it. He later wed Andromeda.

This large constellation is easily found because it is linked to one of the most distinctive constellations in the heavens, Pegasus. The leading star, Alpha Andromedae, or Alpheratz, forms the northeastern corner of the Square of Pegasus.

Andromeda has its fair share of doubles, variables, and clusters, but it is best known for a fuzzy patch of light that looks like a nebula, but isn't, and is clearly visible to the naked eye. M31 in the Messier catalogue, this fuzzy patch used to be called the Great Nebula in Andromeda. However, we now know that it is a neighboring galaxy similar to our own, but considerably bigger and containing many more stars.

The Great Spiral
The Andromeda Galaxy is one of the showpiece objects of the northern heavens. It is a spiral galaxy like our own, but is much bigger, with a diameter of 150,000 light-years. It is in fact the largest galaxy in our Local Group of galaxies, and is well named the Great Spiral.

Of about the fourth magnitude in brightness, the Great Spiral is easily seen near Nu. It lies about 2.3 million light-years away, which makes it easily the farthest naked-eye object in the night sky.

In binoculars the Great Spiral shows as a distinct oval, while larger telescopes bring out individual stars and a dark dust lane. M31 also has two companion galaxies, M32 and M110 (NGC205), both visible in binoculars.

Interesting companions
Among Andromeda's more interesting stars is the second-magnitude Gamma. One of the most beautiful doubles in the night sky, it is readily resolved in small telescopes into a bright golden yellow primary and a bluish-green companion. The companion is also a double, but resolved only in larger telescopes.

An easier double than Gamma is Pi, while a harder one, with yellow components, is 36 Andromedae, located between Eta and Zeta near the border with Pisces.

Among the variable stars in the constellation is the distinctly red Mira-type R Andromedae. It is located close to a trio of fifth-magnitude stars Theta, Rho, and Sigma, north of Alpha in the Square of Pegasus. It comes into binocular range at its brightest (about magnitude 7), but fades to magnitude 14. It is a pulsating red giant star, varying over a period of 409 days.

A green doughnut
Following an arc through Lambda, Kappa, and Iota leads you to NGC7662, one of the brighter planetary nebulae. In small telescopes it looks like a fuzzy greenish star, but larger instruments reveal the doughnut-shaped disk and the very hot central star.

South of Gamma is a trio of stars, 59, 58, and 56. Close to 56, which is a wide double, is a fairly scattered, open cluster, NGC752. Next to Beta (Mirach) is NGC404, an elliptical galaxy which has a fuzzy appearance similar to that of a developing comet.

East of Gamma, in the direction of Algol in the adjoining constellation, is NGC891, which is a classic edge-on spiral, bisected with a dark dust lane. But only larger telescopes will pick this up.

▼ The central part of the Great Spiral, M31. This bright region presents itself in binoculars as an oval. The streak in the picture marks the path of a satellite passing overhead at the time the exposure was being made.

For information about surrounding constellations, see the following pages:

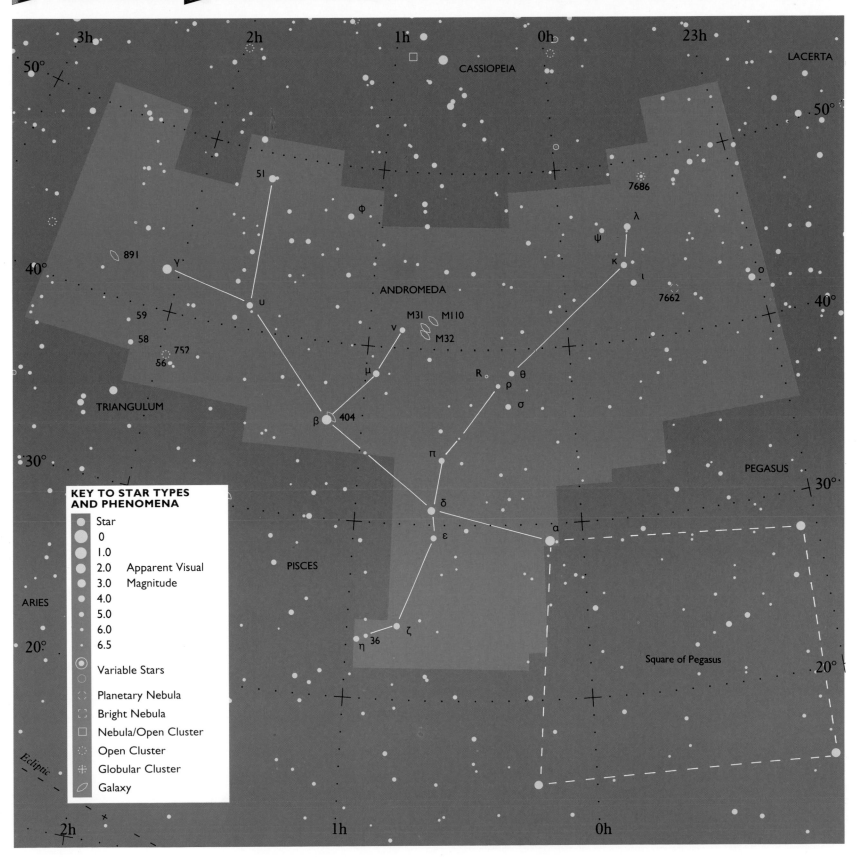

3h
2h
1h
0h
23h

50°
CASSIOPEIA
LACERTA

50°

51

7686

φ
λ

891
γ
ψ
κ
o

40°
ANDROMEDA
ι
7662
40°

59
M31   M110
ν
M32

58
μ
R   θ
752
ρ

56
σ

TRIANGULUM
β   404
π

30°
δ
PEGASUS
30°

PISCES
ε
α

ARIES
Square of Pegasus

20°
ζ
20°

η   36

Ecliptic

2h
1h
0h

**KEY TO STAR TYPES
AND PHENOMENA**

Star
0
1.0
2.0   Apparent Visual
3.0   Magnitude
4.0
5.0
6.0
6.5

Variable Stars

Planetary Nebula

Bright Nebula

Nebula/Open Cluster

Open Cluster

Globular Cluster

Galaxy

CASSIOPEIA

Queen Cassiopeia was the vain wife of King Cepheus and the mother of Andromeda. It was a consequence of the Queen's vanity that Andromeda nearly met an untimely end in the jaws of the sea monster Cetus. Cassiopeia boasted that she was fairer than the Nereids, sea nymphs famed for their beauty. The nymphs were offended and asked Poseidon, who was married to one of their number, to punish the vain Queen. He did so by sending Cetus (see page 78).

With its brightest stars aligned in a distinctive W-shape, Cassiopeia is unmistakable. It is a far northern constellation, located on the opposite side of the Pole Star from Ursa Major. From Canada, the northern United States, and northern Europe, it is circumpolar and always visible.

Cassiopeia sits in the Milky Way, and the region is rich in clusters and dense star fields, which reward leisurely sweeping with binoculars.

The "W" stars

Alpha (Shedir), Beta, and Gamma are almost identical in brightness (about magnitude 2.2). Alpha, the southern-most star of the "W," has a distinctly orange hue, com-

pared with the pure white of Beta. Telescopes will resolve it into a double star, with the orange of the brilliant primary contrasting with the bluish-white of the fainter companion.

Nearby Eta is another fine double star, with golden yellow and purplish components. The central star of the "W," Gamma, varies rather unpredictably. It is a very hot blue giant star that periodically throws off shells of luminous gas, causing its brightness to increase. In the 1930s, it rose to a magnitude of 1.6 to be the brightest star in the constellation, but later dipped to below 3.

Extending the "W" from Delta through Epsilon, you reach Iota. This is an excellent triple star for amateur telescopes.

Interesting variables

Continuing the same line through Iota brings you to RZ Cassiopeiae. This is an eclipsing binary variable star within binocular range. Usually of about the sixth magnitude, it falls nearly to the eighth when it is in eclipse, every 29 hours. The dimming can easily be followed because it takes place over just two hours.

On the other side of the constellation, beyond Beta, is a line of three stars, Tau, Rho, and Sigma. The central star Rho is an irregular variable of unknown type, which fluctuates between about magnitudes 4 and 6 quite unpredictably. The fifth-magnitude flanking stars Tau and Sigma are useful for comparison.

A regular Mira-type variable, R Cassiopeiae, lies south of Rho, though it is not easy to find because of the dense star fields of the Milky Way. At full brightness it reaches the sixth magnitude, but fades to the thirteenth over a period of 431 days.

Fine clusters

Rich star fields lie around Epsilon, Delta, and Gamma, together with a number of clusters. They include the open clusters NGC663, 654, and M103, the brightest. However, the open cluster NGC457, south of Delta and close to Phi, is a more interesting object. It is a compact group of more than 100 stars. Binoculars show it well, and small telescopes will resolve its stars.

Another open cluster worth observing is M52, found, though with some difficulty, by extending a line through Alpha and Beta. It is a cluster of about 100 hot young stars rather like the Pleiades. It is an excellent subject for the small telescope.

▼ One of Cassiopeia's most celebrated objects is all but invisible at optical wavelengths, even in large telescopes. It is Casssiopeia A, the remains of a supernova that happened long ago. At radio wavelengths, as here, Cas-A is one of the brightest objects in the heavens.

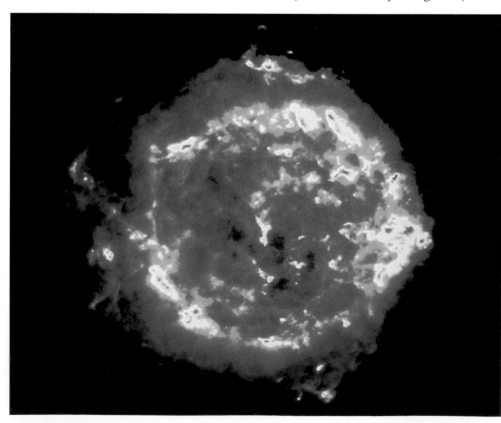

For information about surrounding constellations, see the following pages:

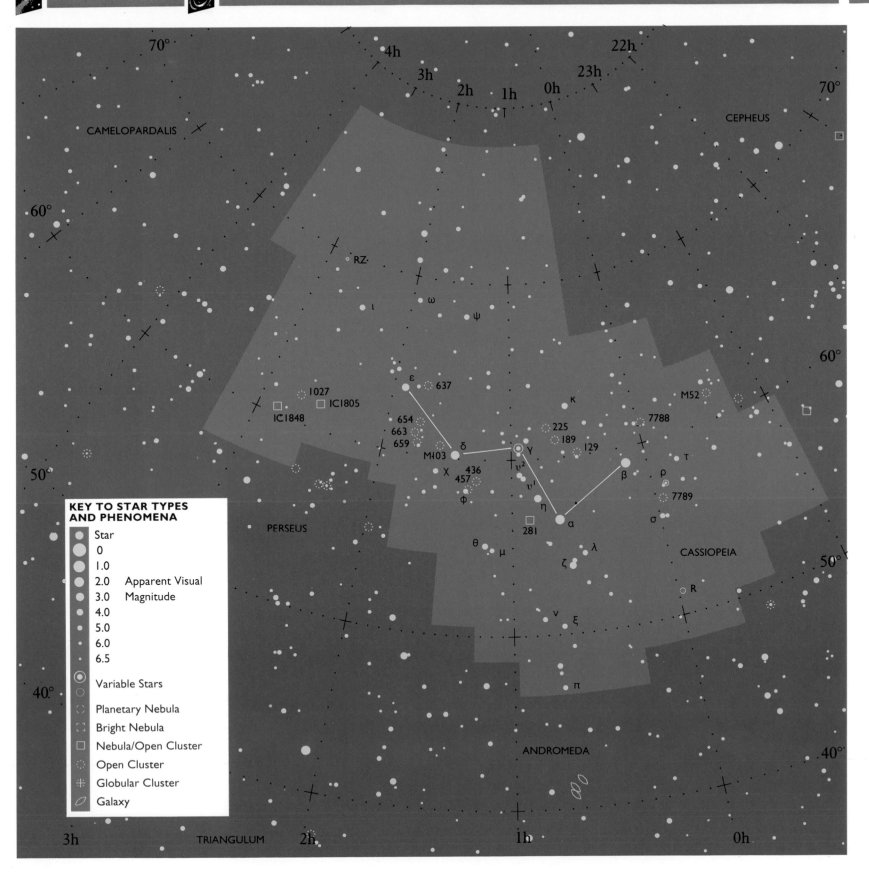

70°
4h
3h
2h
1h
0h
23h
22h
70°
CEPHEUS
CAMELOPARDALIS

60°
60°

RZ
ι
ω
ψ

ε
637
M52
50°
1027
κ
7788
IC1805
225
IC1848
189
654
129
663
δ
γ
τ
659
υ²
β
ρ
M103
χ
436
υ¹
7789
457
σ
φ
η
α
281
CASSIOPEIA
50°
θ
λ
PERSEUS
μ
ζ

R

**KEY TO STAR TYPES
AND PHENOMENA**

Star
0
1.0
2.0   Apparent Visual
3.0   Magnitude
4.0
5.0
6.0
6.5

⊚ Variable Stars
◇ Planetary Nebula
⊡ Bright Nebula
☐ Nebula/Open Cluster
⦂ Open Cluster
✦ Globular Cluster
⬭ Galaxy

ν
ξ

40°
40°

π

ANDROMEDA

3h
TRIANGULUM
2h
1h
0h

CENTAURUS

Centaurus the Centaur was one of the mythical creatures that were half-man, half-horse. Most centaurs were wild creatures, but this particular centaur was different. Named Chiron, he was wise and learned and became a great teacher, especially of hunting, healing, and music. Despite his medical skills, he could not heal himself when he was accidentally hit by one of Hercules' poisoned arrows.

Centaurus is one of the largest constellations in the heavens and one of the most spectacular. It has its "feet" in the Milky Way and envelops the distinctive Crux, the Southern Cross.

Centaurus boasts the third brightest star in the night sky, the nearest star, a wealth of open and globular clusters and galaxies, and it contains one of the most intense radio sources in the sky.

The Alpha Centauri system

The leading star Alpha, sometimes called Rigil Kent, has a magnitude of −0.3 and is outshone in the heavens only by Sirius and Canopus. Alpha Centauri is a double star, separated in small telescopes into yellow components. It lies just 4.3 light-years away, less than half the distance of Sirius.

When Alpha is viewed in large telescopes, a third faint star comes into view. It is a red dwarf of the eleventh magnitude, and it is closer even than Alpha. Indeed, it is the nearest star to the Earth and is appropriately called Proxima Centauri.

Beta, also called Agena, has a magnitude of 0.6. It forms a prominent pair with Alpha, and the two are called the Pointers because they point to the Southern Cross. Being bluish-white, Beta contrasts with yellow Alpha. Though side by side in the heavens, the two are literally light-years apart, more than 450 in fact.

Between the Pointers is the Mira-type variable R. It is an easy binocular subject at its maximum, when it is at the fifth magnitude, but fades to the eleventh at minimum over a period of 547 days.

Outstanding Omega

Following a line from Beta through Epsilon to an equal distance beyond brings you to yet another glorious feature of the southern skies. It looks like a fuzzy fourth-magnitude star labeled Omega.

However, Omega Centauri is not a star at all, but a grouping of hundreds of thousands of stars, maybe as many as a million! It is in fact the finest example of a globular cluster there is. It is readily visible to the naked eye, and even binoculars will begin to resolve its stars.

Another globular cluster worth investigating in binoculars and a small telescope is NGC5286, just north of Epsilon. One of the finest open clusters in the constellation is NGC3766, near Lambda on the opposite side of the Southern Cross from Alpha and Beta.

Crux, the Southern Cross

This, the smallest constellation, is circumpolar in New Zealand and much of Australia. Four main stars make up the Cross, Alpha (Acrux), Beta, Gamma, and Delta. Warm reddish Gamma makes a nice contrast with the other bluish-white stars. Alpha (magnitude 0.8) is a double, as are Beta and Gamma and the pair Theta.

Near Beta is Crux's finest gem, a beautiful colored open cluster nicknamed the Jewel Box, but officially termed Kappa Crucis or NGC4755. It is a glittering spectacle as its name implies, featuring a prominent red supergiant that shines like a ruby.

Kappa Crucis lies at the edge of a dark "hole" in the Milky Way star fields, appropriately named the Coal Sack. It is, in fact, a large dark nebula.

► Stars beyond number cluster together in one of the finest sights in Centaurus, Omega Centauri. It is the largest, brightest, and richest globular cluster in the heavens, occupying about the same area of the sky as the Full Moon. It lies about 17,000 light-years away.

For information about surrounding constellations, see the following pages:

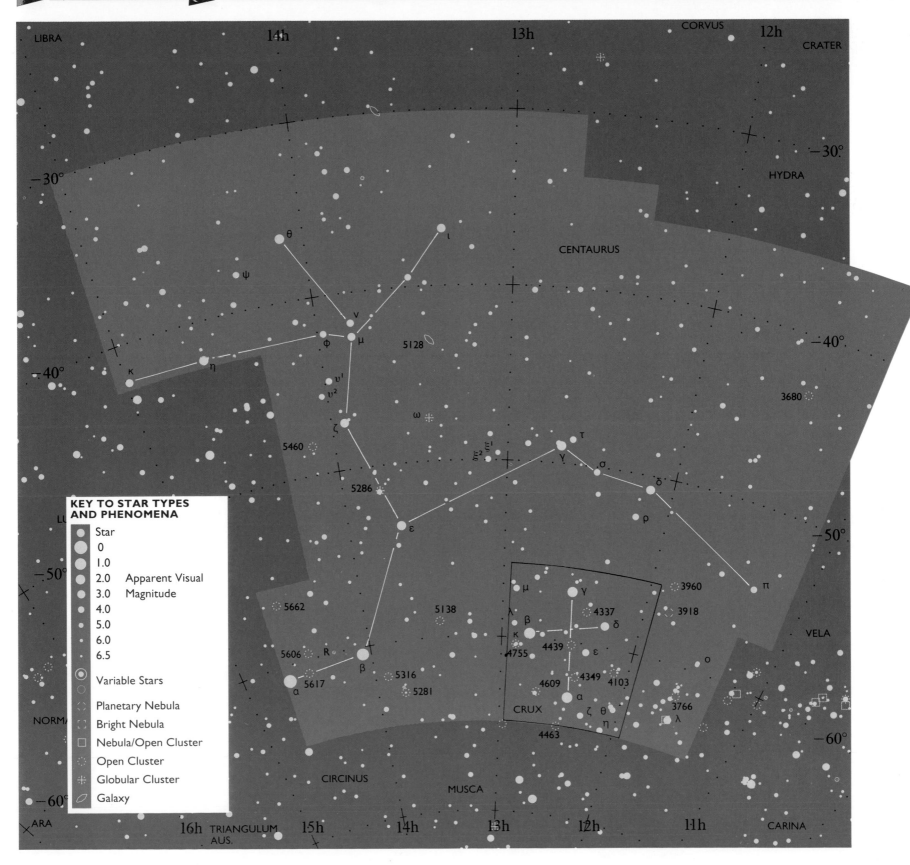

LIBRA

14h

13h

CORVUS

12h

CRATER

−30°

HYDRA

CENTAURUS

−30°

θ

ι

ψ

−40°

ν

5128

μ

Φ

υ¹

υ²

η

κ

ζ

5460

ω

π²
π¹

τ

γ

σ

δ

5286

ρ

−40°

ε

π

−50°

**KEY TO STAR TYPES
AND PHENOMENA**

Star

0

1.0

2.0   Apparent Visual

3.0   Magnitude

4.0

5.0

6.0

6.5

⊙ Variable Stars

◇ Planetary Nebula

⊡ Bright Nebula

▫ Nebula/Open Cluster

○ Open Cluster

✻ Globular Cluster

⬭ Galaxy

−50°

5662

5138

3960

μ

γ

λ

4337

3918

β

δ

κ

4439

ε

4755

VELA

5606

R

5316

4609

4349

4103

α

5617

β

5281

α

3766

CRUX

ζ θ

3766

λ

η

4463

−60°

NORMA

CIRCINUS

MUSCA

−60°

ARA

16h

15h

14h

13h

12h

11h

CARINA

TRIANGULUM
AUS.

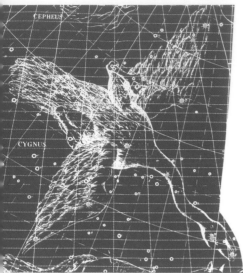

CYGNUS

Cygnus the Swan is supposed to be Zeus (Jupiter), king of the gods, in disguise. Zeus assumed this form when visiting one of his many lady loves, Leda, Queen of Sparta. The result of the liaison was an egg, out of which hatched the twins Castor and Pollux and also Helen, who would become the beautiful Helen of Troy, with the "face that launched a thousand ships." Pollux and Helen were both Zeus's children, but Castor was not; he was the son of the King of Sparta.

▼ The delicate Veil Nebula (NGC6992), which is part of the extensive supernova remnant known as the Cygnus Loop. The name is a good one for these wispy threads of glowing gas, as is the alternative name, the Cirrus Nebula.

Cygnus is a beautiful northern constellation and one of the few to resemble the figure it is named for. The Swan is depicted flying south along the Milky Way. Cygnus's distinctive cross shape gives it the alternative name of the Northern Cross.

Cygnus provides an astronomical feast. It features many fine doubles and variables. Being largely in the Milky Way, it also abounds in star fields and clusters, which put on a magnificent display in binoculars. Larger telescopes reveal magnificent nebulae and diaphanous wisps of glowing gas.

Dazzling Deneb
The leading star Alpha, or Deneb, in the Swan's tail, is of the first magnitude. It is exceptionally luminous, with the light output of more than 60,000 Suns. It lies very much farther away than all the other bright stars, at a distance of more than 1,800 light-years. If the brightest star in the sky, Sirius, were at that distance, we would hardly notice it!

Deneb is one of the three bright stars that make up the Summer Triangle. The others are Altair in Aquila and Vega in Lyra.

Lovely doubles
Beta, or Albireo, is at the opposite end of Cygnus from Deneb and is much fainter (third magnitude). But viewed even through a small telescope, it is a gem of a double, probably the loveliest in the sky. Its blue and yellow components always appear clear and bright.

Delta and Gamma are also double, as is 61, under the Swan's right "wing." 61 Cygni has a greater claim to fame than just being a double. It was the first star to have its distance measured accurately. The German astronomer Friedrich Bessel worked out the distance, by the parallax method, in 1838. It proved to be over 65¼ million million miles (105 million million kilometers). That the stars were so remote caused a sensation.

Red variables
Cygnus contains several Mira-type variables, visible in binoculars at maximum brightness and worth viewing if only for their intense red color. One of the most interesting is Chi, located in the Swan's neck. At maximum, Chi Cygni rivals the steady and nearby Eta (magnitude 3.6) in brightness. At minimum, it fades to below magnitude 14 over a period of 407 days, thus displaying one of the greatest changes in brightness known among variables.

Other Mira variables include U and R Cygni which come within binocular range near maximum, but fade way beyond it for much of the time. U is east of the star 32, which is itself north of the pair Omicron 1 and 2 under the Swan's left wing. R is located close to Theta farther out.

W Cygni, on the opposite side of the constellation close to the fourth-magnitude Rho, is a semi-regular variable with a period of about four months. Its change in brightness, from the fifth to the eighth magnitude, keeps it within range of binoculars throughout its cycle.

Deep-sky revelations
In all, there are up to 30 open clusters in the constellation, including two Messier clusters. M29 is located just south of Gamma; and M39 is found in the tail end of the constellation, forming a right-angled triangle with Pi-2 and Rho. M39 is the larger of the two, with more than 20 stars covering nearly half a Moon diameter.

Between Deneb and the much fainter Xi, the Milky Way noticeably brightens. Viewed through binoculars or a telescope, the region is revealed as a large nebula (NGC7000). Long-exposure photographs show a distinctive shape that gives it the name of the North America Nebula.

South of Epsilon in the Swan's right wing is a fourth-magnitude star 52. When the region is viewed in large binoculars or a telescope, filaments of glowing gas show up. And long-exposure photographs reveal a delicate looping tracery known as the Cygnus Loop. It is a supernova remnant, the remains of a supernova explosion that happened maybe 30,000 years ago. We know a bright part of the Loop as the Veil Nebula, sometimes called the Cirrus Nebula.

For information about surrounding constellations, see the following pages:

Cassiopeia 80	Cepheus 40
Draco 40	Lyra 56
Pegasus 62	

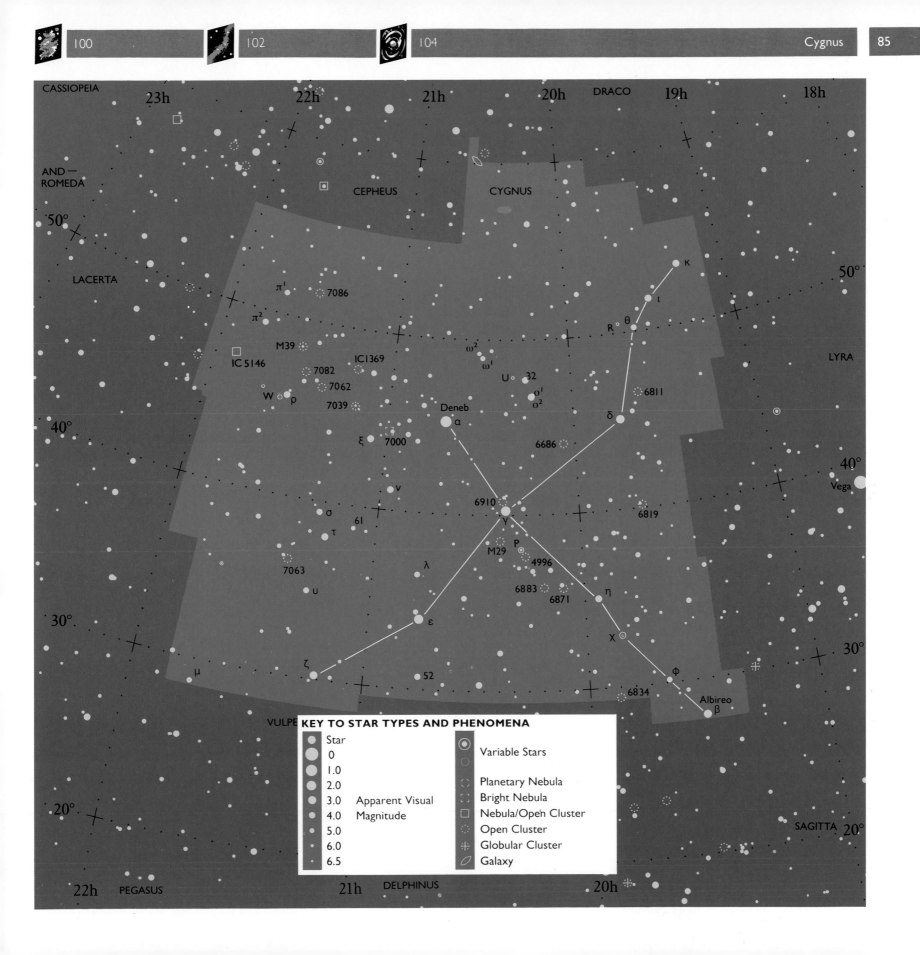

CASSIOPEIA

23h 22h 21h 20h DRACO 19h 18h

AND — ROMEDA

CEPHEUS CYGNUS

50° 50°

LACERTA κ

π¹ ι
7086

π² R θ
LYRA

M39 ω²

IC 5146 IC 1369 ω¹ 6811

7082 U 32

W ρ 7062 o¹
7039 o²

40° Deneb δ 40°
α 6686

ξ 7000

ν 6910

σ 6819
61 γ

τ P
M29 4996

7063 λ 6883 η
u 6871

χ
ε φ
30° 30°

ζ 6834
μ 52 Albireo
β

VULPE

KEY TO STAR TYPES AND PHENOMENA

Star
0
1.0
2.0
3.0   Apparent Visual
4.0   Magnitude
5.0
6.0
6.5

Variable Stars

Planetary Nebula
Bright Nebula
Nebula/Open Cluster
Open Cluster
Globular Cluster
Galaxy

20° SAGITTA 20°

22h PEGASUS 21h DELPHINUS 20h

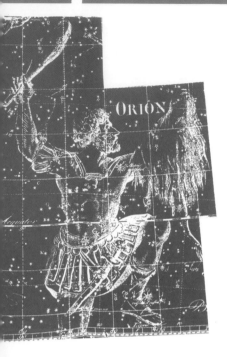

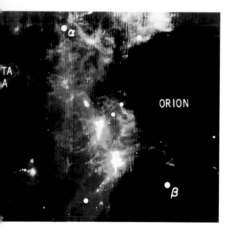

▲ Our telescopes reveal that Orion contains many fine nebulae, notably M42. But the infrared astronomy satellite IRAS revealed that gas and dust pervade practically the whole of the constellation. This IRAS image highlights this vast region of interstellar matter. Alpha (Betelgeux) is pinpointed.

ORION

In mythology, Orion was a mighty hunter. He was the son of the sea god Poseidon, who gave him the power to walk on the sea. Orion features in many tales. In one he fell in love with seven sisters, the Pleiades, and pursued them, as he does still in the heavens. Different stories tell of how Orion died. One tells of Orion's boast that he could kill any creature alive, which caused the Earth to open. Out came a scorpion, which promptly stung him to death.

The celestial equator passes through this magnificent constellation, which is therefore equally familiar to sky-watchers in both hemispheres.

"Magnificent" is certainly the right word to describe Orion, whose main pattern of bright stars makes it one of the most readily recognizable of all the constellations. Also, its outline really does look like the figure it is meant to represent, the mighty hunter with club upraised and shield at the ready.

Twin giants
Two stars vie with each other to be the finest in the constellation, Alpha, or Betelgeux (sometimes spelt Betelgeuse and pronounced "beetle-juice"), and Beta, or Rigel.

Betelgeux is supergigantic; if it were where the Sun is, its globe would reach beyond the Earth's orbit nearly to Mars. It is obviously red and variable. At maximum brightness, it reaches a magnitude of 0.5, but can fade below 1 over an irregular period of about five to six years.

Rigel, which is diametrically opposite Betelgeux in the main star pattern, is actually brighter, shining steadily at magnitude 0.1. It too differs in color, being bluish-white. It is an intensely hot star with the energy output of more than 50,000 Suns. Larger telescopes will show it to be a double.

The other bright stars in Orion are also named: Gamma – Bellatrix, Delta – Mintaka, Epsilon – Alnilam, Zeta – Alnitak, and Kappa – Saiph.

Great nebulae
Apart from the main star pattern, the most conspicuous feature of the constellation to the naked eye is a misty patch below the three stars, Delta, Epsilon, and Zeta, which form Orion's Belt. Binoculars show it to be a bright nebula, M42, aptly named the Great Nebula in Orion. Long-exposure photographs of the nebula are needed to bring out its staggering beauty, which few other heavenly sights can match.

A much smaller, nearly round nebula, M43, lies close to the Great Nebula. There is a small reddish nebula close to Zeta, identified as NGC2024, and a slightly brighter and more bluish one farther north, M78.

South of Zeta is the glowing nebula IC434, which has obscuring dark matter silhouetted against it in the shape of a horse's head. This Horsehead Nebula, also called Barnard 33, is the most distinctive dark nebula in the heavens. Unfortunately, it is beyond the reach of smaller telescopes.

The nebulae highlighted above are really the "tip of the iceberg" because virtually the whole constellation sits in a tenuous gas cloud. It is this vast reservoir of interstellar matter that spawned hot white stars like Rigel and those in Orion's Belt.

Multiple choice
Orion is rich, too, in double and multiple stars. Pride of place must go to the star Theta Orionis, which is embedded in the Great Nebula. Called the Trapezium after the arrangement of its four components, it can be separated in small telescopes. It is the light from these that illuminates the Nebula. Iota, at the southern edge of the Nebula, is also double.

Going northward, Sigma appears double in binoculars, but small telescopes reveal it as a triple, with white, bluish, and reddish components. Larger instruments separate a fourth component. In Orion's Belt, Zeta and Delta are double. So is Lambda, the third-magnitude star marking Orion's "eye."

And Orion has still more to offer, in the way of variables, open clusters, and planetary nebulae. Perhaps the easiest variable to find is the very red W, at the southern end of the arc of the six Pi stars that form Orion's Shield. W is a long-period variable of the Mira type, which varies between about the sixth and eighth magnitudes over a period of 212 days and is always within binocular range.

The open cluster NGC2169 is the brightest in the constellation at about the eighth magnitude. It forms an equilateral triangle with Xi and Nu in Orion's upraised right hand. Going farther north, NGC2174/5 is an attractive combination open cluster and nebula, in line with Chi 1 and 2.

For information about surrounding constellations, see the following pages:

Eridanus 66	Gemini 46
Lepus 45	Monoceros 47
Taurus 92	

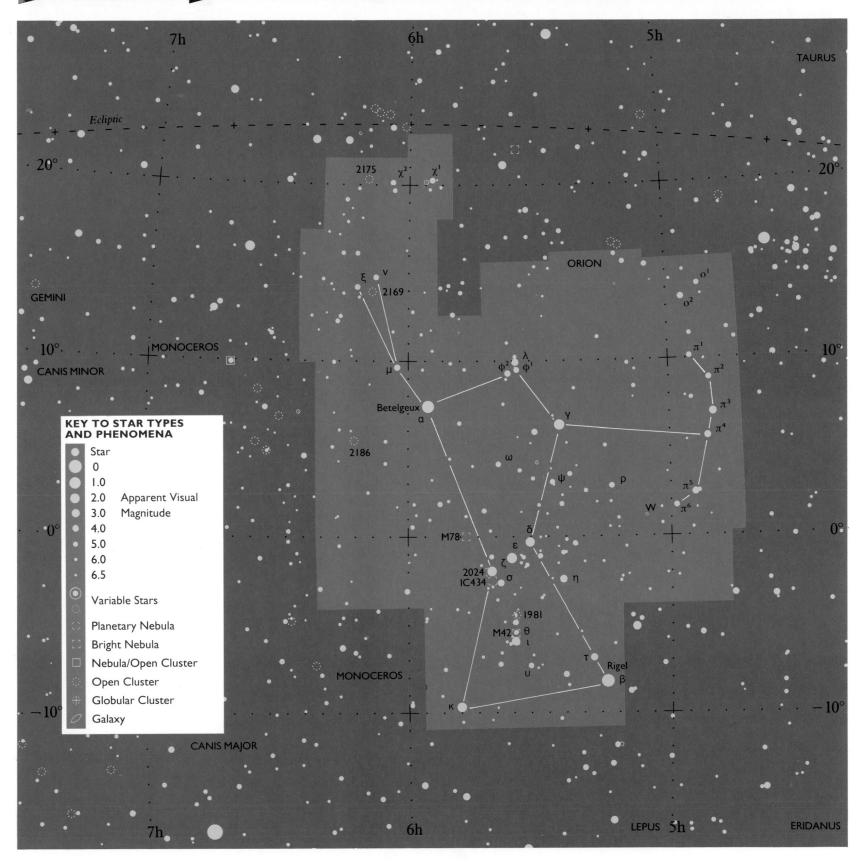

KEY TO STAR TYPES AND PHENOMENA

Star
0
1.0
2.0   Apparent Visual
3.0   Magnitude
4.0
5.0
6.0
6.5

Variable Stars

Planetary Nebula

Bright Nebula

Nebula/Open Cluster

Open Cluster

Globular Cluster

Galaxy

SAGITTARIUS

Sagittarius the Archer is depicted as a half-man, half-horse centaur, but one of the warlike ones (unlike Centaurus, page 82). He has his bow drawn, with the arrow pointing to Antares, the Scorpion's heart.

In early times, Sagittarius was thought of rather as a satyr, a two-legged creature that was half-man, half-goat. He was identified with Crotus, the inventor of archery, whose father was the pipe-playing god Pan.

Sagittarius is one of the richest constellations in the heavens, spanning the Milky Way. The Milky Way here is magnificent, studded with clusters and beautiful nebulae and brilliant with dense star clouds. The densest mark the direction of the center of our Galaxy some 30,000 light-years away. Swept with binoculars, this part of the night sky takes the breath away.

Alas for most skywatchers in the Northern Hemisphere, Sagittarius never rises high enough to be fully appreciated. How they envy astronomers down under!

A Messier feast
The constellation has a number of double and variable stars (such as the Cepheid variable W, close to Gamma). However, its chief attraction lies in its deep-sky objects. Charles Messier catalogued more of them in Sagittarius than in any other constellation, including some of the best-known nebulae and many globular clusters.

M8, the Lagoon Nebula, is one of the brightest of the Messier nebulae. Of the fifth magnitude, it can be glimpsed with the naked eye, making a triangle with Lambda and Mu. It is a fine object in binoculars and a

small telescope, but as always with nebulae, only long-exposure photographs bring out its full beauty. Such pictures show up dark gas lanes and many tiny black condensations known as Bok globules, which are probably on the way to becoming stars.

Just to the north of M8 is an equally famous nebula, M20, the Trifid. This is visible in binoculars as a small misty patch. Telescopes will reveal its three dark dust lanes, which give the nebula its name.

The third major nebula in Sagittarius, M17, is located at the northern border of the constellation on the edge of the Milky Way. It is probably best found by following a line between Gamma Scuti in the neighboring constellation and Mu. M17 has roughly the shape of the Greek letter omega, and so is usually called the Omega Nebula. It has the alternative names of the Swan and the Horseshoe Nebula.

Cluster collection
Immediately south of the Omega Nebula is the open cluster M18 and south of this M24. M24 is neither a nebula nor a cluster, but a general brightening of the Milky Way otherwise known as the Small Sagittarius Star Cloud.

Making a triangle with M24 and Mu is a good binocular open cluster M23, a collection of over 100 stars of up to the ninth magnitude. Farther south, near the Trifid Nebula, is a similar object, M21.

Great globulars
Sagittarius boasts more than 20 globular clusters, about one-tenth of the number known. That it contains so many is not really surprising, because globulars tend to congregate around the centers of galaxies.

The top globular is undoubtedly M22, which forms a triangle with Lambda and Phi. M22 is a dense grouping of literally millions of stars. It rates the sixth magnitude and is just visible to the naked eye. It looks great in binoculars and small telescopes, which begin to resolve its outer stars.

Other notable globulars include M54 and M55. M54 is located close to Zeta, while M55 is at about the same declination some way to the east. M55 is a much looser condensation of stars, counted in thousands rather than millions like M22.

▲ One of the dense star clouds in Sagittarius, looking toward the center of the Galaxy. This region is a delight for binocular observers.

▶ The Lagoon Nebula (M8) is named because of the shape of the dark lane running through it. It is a vast object, maybe as much as 100 light-years across. It lies about 5,000 light-years away.

For information about surrounding constellations, see the following pages:

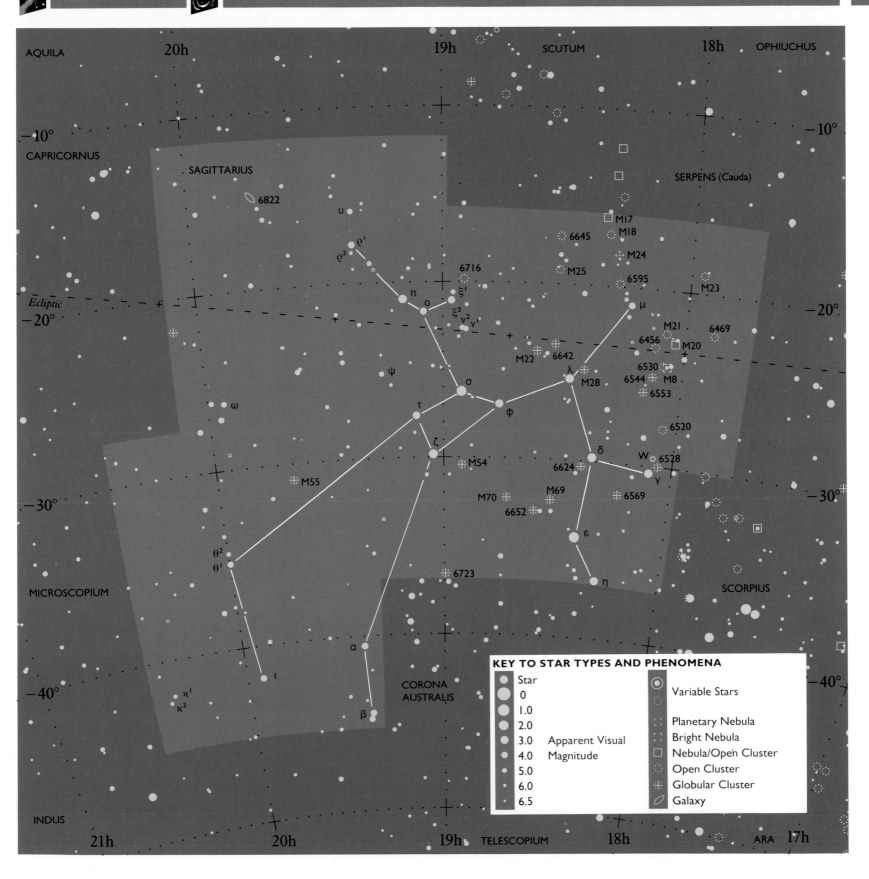

AQUILA 20h 19h SCUTUM 18h OPHIUCHUS

−10° −10°

CAPRICORNUS

SAGITTARIUS SERPENS (Cauda)

6822

υ

ϱ¹
ϱ²

6716 M17

6645 M18

π M24

ξ¹ 6595
ξ² M25 M23
ν² ν¹ μ

Ecliptic

−20° −20°

M21 6469

6456 M20

ψ M22 6642 6530

λ 6544 M8

σ M28 6553

ω

τ 6520

ζ δ W 6528

M54 6624 γ

M55 6569

θ² M70 M69

θ¹ 6652

−30° ε −30°

MICROSCOPIUM SCORPIUS

6723

η

α

CORONA
AUSTRALIS

ι

−40° −40°

χ¹

χ²

β

INDUS

21h 20h 19h TELESCOPIUM 18h ARA 17h

KEY TO STAR TYPES AND PHENOMENA

Star		Variable Stars
0		Planetary Nebula
1.0		Bright Nebula
2.0		Nebula/Open Cluster
3.0	Apparent Visual	Open Cluster
4.0	Magnitude	Globular Cluster
5.0		Galaxy
6.0		
6.5		

SCORPIUS

Scorpius the Scorpion was the creature that stung to death the mighty hunter Orion. Orion and Scorpius were therefore placed on opposite sides of the heavens, so that Orion disappears as the Scorpion rises. In the sky, however, Scorpius itself is threatened by Sagittarius the Archer, whose arrow is pointing at the Scorpion's heart, marked by Antares. The constellation that is now Libra once formed the Scorpion's front claws.

▲ The whole constellation of Scorpius, photographed by the author in the Australian Outback under perfect seeing conditions. It was taken by a tripod-mounted 35-mm camera with standard lens. The exposure time was about half a minute on Ektachrome 400 film "pushed" to 800. It demonstrates that reasonable night-sky pictures can be obtained even with simple photographic equipment.

Scorpius is another of those southern constellations that are the envy of northern astronomers. It is relatively small, but brilliant, boasting no less than 11 stars above the third magnitude. Also, practically all of the constellation lies within the Milky Way, and in one of the richest parts of it, too, where open and globular clusters abound.

The name given to the constellation is, for a change, appropriate. Only the slightest imagination is required to picture a scorpion poised to strike, with head in the north of the constellation and its curved tail in the south.

Rival of Mars

Scorpius's brightest star is Alpha, or Antares, which marks the heart of the Scorpion. Its name means "rival of Mars," alluding to its resemblance to the Red Planet. It is perhaps the reddest of the bright stars, more so than Betelgeux, for example.

Like Betelgeux, Antares is a red supergiant, but it is even bigger, probably as much as 700 times the size of the Sun. If it were where the Sun is, its surface would stretch beyond the orbit of Mars up to the asteroid belt.

Being so large, Antares is unstable and pulsates, changing in brightness slightly as it does so and averaging about magnitude 1.

The Scorpion's "head"

An arc of four stars, Nu, Beta, Delta, and Pi, mark the Scorpion's "head." In the present constellation, the claws are missing, the claw stars now forming part of the neighboring constellation Libra.

Of the "head" stars, Nu appears double in binoculars, and telescopes show each component itself to be double, making Nu a quadruple system. Second-magnitude Beta is also a fine double star with bluish-white components. Nearby Omega is an easy binocular double.

Moving south along the "backbone" of bright stars, Sigma is double (as, incidentally, is Antares). Mu and Zeta, at the base of the "tail," are wide, easy doubles. Zeta 1, the brighter component, is reddish-orange; Zeta 2 is bluish-white. In binoculars or a small telescope, the region around Zeta is fascinating. It includes the third-magnitude open cluster NGC6231 composed, like the Pleiades, of young hot stars and plainly visible to the naked eye.

Sting in the tail

The tail region of Scorpius is also a delight for naked-eye skywatcher, binocular viewer, and small-telescope observer alike. Lambda, or Shaula, at magnitude 1.6, marks the Scorpion's sting. It is only fractionally brighter than Theta farther south.

M7 and M6 are bright open clusters visible to the naked eye north of the tail. Both can be resolved into stars in binoculars. The third-magnitude M7 is the brighter and larger of the two, covering an area of sky larger than that of the full moon. M6 is also known as the Butterfly Cluster, because of the "open-wing" arrangement of its principal stars.

Other Messier objects in the constellation include the sixth-magnitude M4 right next to Antares, a fine globular cluster for small telescopes. M80 is a somewhat fainter globular roughly midway between Antares and Beta.

For information about surrounding constellations, see the following pages:

Ara	42	Libra	53
Lupus	53	Ophiuchus	56
Sagittarius	88	Serpens Cauda	57

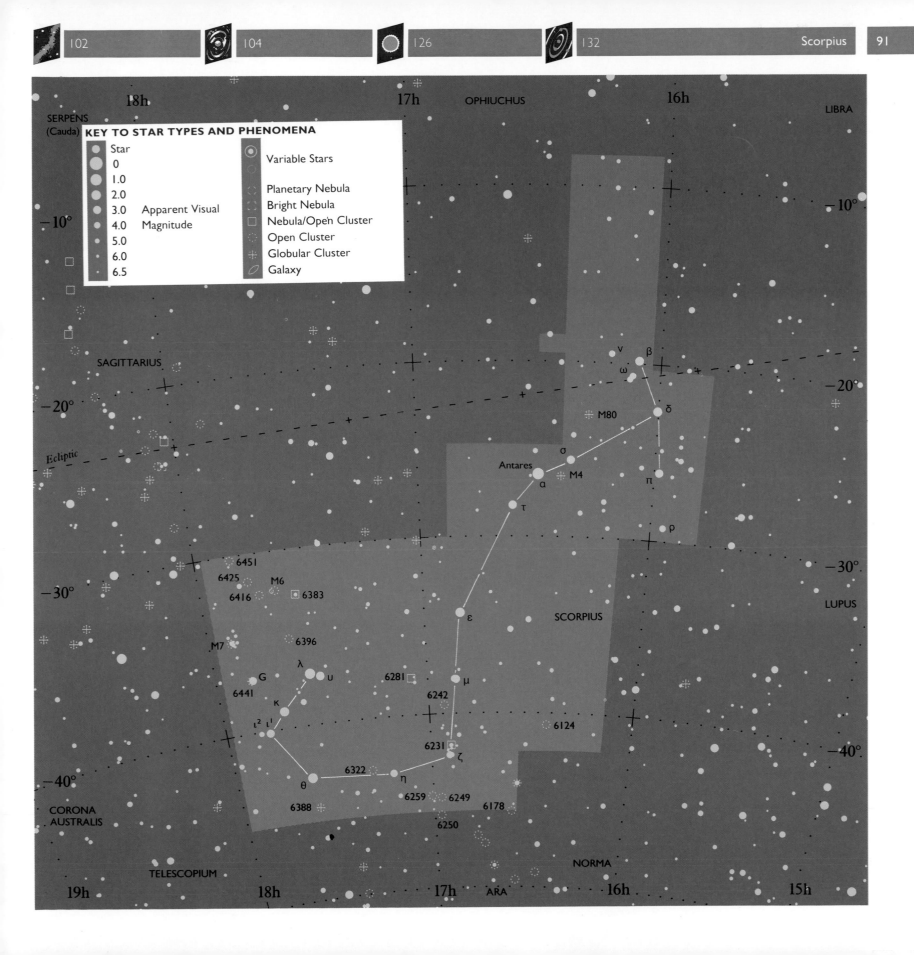

SERPENS
(Cauda)

KEY TO STAR TYPES AND PHENOMENA

Star
0
1.0
2.0
3.0   Apparent Visual
4.0   Magnitude
5.0
6.0
6.5

Variable Stars

Planetary Nebula
Bright Nebula
Nebula/Open Cluster
Open Cluster
Globular Cluster
Galaxy

18h

17h   OPHIUCHUS   16h   LIBRA

−10°

SAGITTARIUS

−20°

Ecliptic

ν   β
ω
δ
M80
σ
Antares   M4
α
π
τ
ρ

SCORPIUS   LUPUS

−10°

−20°

−30°

ε

6451
6425   M6
6416   6383

6396

M7
G   λ
u
6441   6281
κ   6242
ι² ι¹   μ
6124
6231
ζ
6322   η
θ   6259 6249
6259 6249   6178
6388   6250

CORONA
AUSTRALIS

TELESCOPIUM

19h   18h   17h   ARA   16h   15h

NORMA

−30°

−40°

−40°

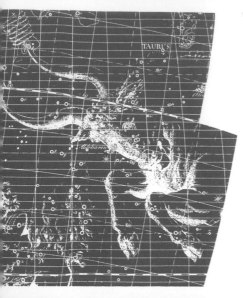

TAURUS

Taurus the Bull was one of the many disguises adopted by king-of-the-gods Zeus on his amorous adventures. Zeus assumed the form of a magnificent snow-white beast with shining horns when he set out to woo the fair Europa. He encouraged her to climb on his back, then carried her off to the island of Crete. There he revealed his true self. One of their children was King Minos, who built the palace of Knossos.

This large constellation adjoins Orion, and as far as stars are concerned, it is faint by comparison, but it boasts a fine leading star which marks the "eye" of the Bull, two of the brightest open clusters in the heavens and the best-known supernova remnant.

The leading star is Alpha, or Aldebaran, which is easily found by following a line through Orion's Belt toward the northwest. Aldebaran is unmistakable both because of its brightness (magnitude 0.8) and its lovely orange color. It is similar in brightness and appearance to Betelgeux in Orion, not far away.

Beta, or Al Nath, is the second brightest star (magnitude 1.6), and marks the tip of one of the Bull's horns. Third-magnitude Zeta marks the tip of the other. Both Beta and Zeta lie on the edge of the Milky Way.

Among the other stars, Lambda is worth watching because it is an eclipsing binary of the Algol type. Every four days, its brightness dips momentarily from the third to the fourth magnitude as the dim star of the pair passes in front of the bright one.

The Hyades

Close to Aldebaran lies a scattered group of stars which form a distinctive V-shape and which are clearly visible to the naked eye. Known as the Hyades, they form one of the finest open clusters in the night sky.

The brightest star in the Hyades is Theta, a naked-eye double consisting of a white primary of magnitude 3.4 and a slightly fainter orange companion. Binoculars show the color difference well. Several of the other stars in the cluster are of similar brightness, including Epsilon, Delta, and Gamma. Delta forms a double with a fainter star 64 Tauri. Sigma is another double close to Aldebaran.

Under good viewing conditions, as many as 20 stars can be detected in the Hyades with the naked eye. This is about one-tenth of the total number of stars in the cluster. On average, the Hyades lies about 130 light-years away, making it one of the nearest open clusters. Aldebaran

itself is not part of the group, being only about half the distance away.

Seven Sisters

Northwest of the Hyades lies an even more spectacular cluster, M45 or the Pleiades. It is also called the Seven Sisters, because keen-sighted people should be able to spot at least seven of its brightest stars with the naked eye. Binoculars and telescopes reveal more and more stars, which probably number in total more than 300.

Compared with the Hyades, the Pleiades is much more concentrated and thus more brilliant, shining nearly as brightly as a first-magnitude star. It differs from the Hyades in several other ways. It is much farther away, at about 400 light-years, and it is much younger. Whereas the Hyades stars are on average about 400 million years old, those in the Pleiades are youngsters of 10–20 million years old. They are hot and white, and surrounded by nebulosity which becomes visible in photographs.

The brightest Pleiades star is Alcyone (magnitude 2.9), and in descending order of brightness come the other six of the "Seven Sisters": Atlas, Electra, Maia, Merope, Taygete, and Pleione.

The Crab

Close by Zeta, at the tip of the Bull's southern horn, is another of Taurus's great sights, M1, the Crab Nebula. It is a supernova remnant, an expanding cloud of writhing gas that resulted from a supernova explosion. It is one of the few whose precise age we know. The supernova was witnessed in July 1054 by Chinese astronomers, who recorded that it was so bright that it was visible in daylight for several weeks.

The Crab is just visible in powerful binoculars, and in small telescopes appears as an oval glow. Larger telescopes are required for closer study. They bring out bright interlacing filaments which appear red in long-exposure photographs.

The nebula was named, incidentally, in 1844 by the English astronomer, the third Earl of Rosse, but it had been discovered over a century earlier. It was the first sight of the nebula by the French astronomer Charles Messier in 1758 that inspired him to begin his historic catalog of nebulous objects. This explains why the Crab Nebula is M1.

▲ This is the famous Crab Nebula, the still-expanding cloud of gas that was blasted into space from a supernova in 1054. At the heart of the Crab is a tiny rotating neutron star, or pulsar, which flashes pulses of energy toward us 30 times a second.

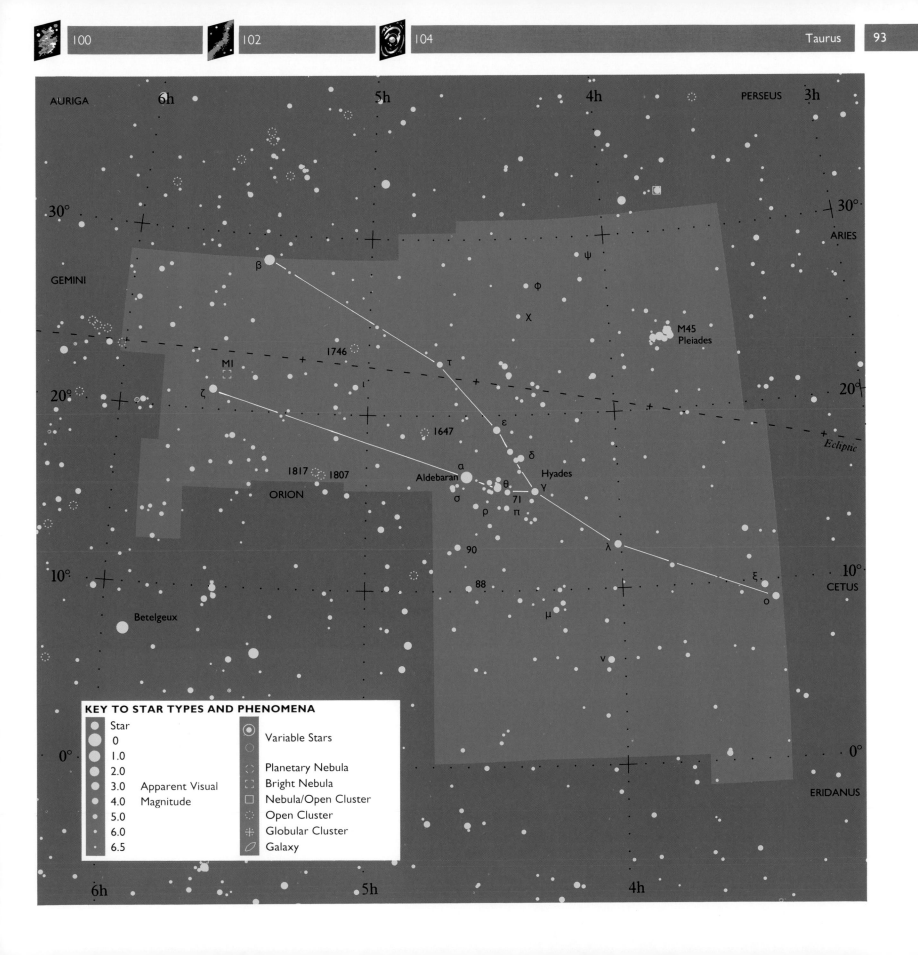

AURIGA   6h   5h   4h   PERSEUS   3h

30°   30°

GEMINI   ARIES

β   ψ

φ

χ   M45
                                                          Pleiades

MI   1746   τ   Ecliptic
ζ
20°   ε   20°
                        1647   δ

1817 1807   α   Hyades
              ORION   Aldebaran   θ   γ
                                      σ   71
                                          ρ   π
                                                      λ
                90   ξ
10°   CETUS   10°
88   o

Betelgeux   μ

                                          ν

0°   0°

KEY TO STAR TYPES AND PHENOMENA

Star
0
1.0
2.0
3.0   Apparent Visual
4.0   Magnitude
5.0
6.0
6.5

⊙ Variable Stars
○

◇ Planetary Nebula
▢ Bright Nebula
□ Nebula/Open Cluster
∴ Open Cluster
✤ Globular Cluster
⬭ Galaxy

ERIDANUS

6h   5h   4h

URSA MAJOR

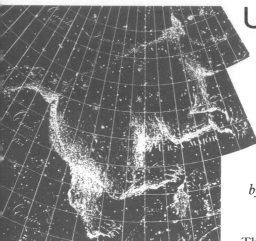

Ursa Major the Great Bear is usually associated with the huntress Callisto, a favorite of Artemis, the goddess of hunting. One day Zeus, cunningly assuming the guise of Artemis, lay with Callisto and made her pregnant. She bore a son, Arcas. Zeus's wife Hera was not amused and turned Callisto into a bear. Later, Zeus placed Callisto in the heavens as the Great Bear, to prevent her from being killed by her son, who came upon her when hunting.

The third largest constellation, Ursa Major occupies a vast region of far northern skies. It is circumpolar for northern Europe, Canada, and the northern United States. Even though it does not boast any stars of the first magnitude, it is the best-known constellation in the Northern Hemisphere. This is because of the distinctive pattern made by seven bright second- and third-magnitude stars which stands out clearly in an otherwise relatively dark area of sky.

This pattern is known as the Big Dipper because of its resemblance to a ladle. It is called the Plough in Europe because it also resembles the handle and plow blade of an old horse-drawn plow.

It should be emphasized that the Big Dipper itself is not a constellation, but represents just the "tail" and "hindquarters" of the Bear. It occupies only a small part of the whole constellation of Ursa Major.

Following the Big Dipper
As might be expected in such a prominent pattern, the stars all have names. Alpha is named Dubhe, Beta is Merak, Gamma is Phad, Delta is Megrez, Epsilon is Alioth, Zeta is Mizar, and Eta is Alkaid.

The Alpha-to-Eta designation in this instance does not give the order of brightness of the stars. In fact, Epsilon vies with Alpha to be the brightest star, both having a magnitude of 1.8. Eta runs them a close third, followed by Zeta. Delta is the dimmest of the seven, at magnitude 3.3.

Perhaps the most interesting of the Big Dipper stars is Mizar (Zeta), marking the bend in the ladle's handle. Mizar is a complex multiple star. To begin with, it is a naked-eye double, whose companion is a fourth-magnitude star named Alchor. Seen through a small telescope, Mizar itself is seen to be double, and examination of the spectra of the two components reveals that they are both close binary systems.

▲ M101 is a striking galaxy of the Sc type, with open and well-defined spiral arms. Its apt popular name is the Pinwheel Galaxy. It is one of the largest spirals known, and is twice as big across as our own Galaxy.

Pointing north
Dubhe (Alpha) differs from the other stars in the Big Dipper, which are white, by having a pronounced orange hue. It and Merak (Beta) are known as the Pointers, because a line drawn northward through them points to Polaris, the Pole Star.

For this reason, the Big Dipper has been a boon to navigators in the Northern Hemisphere for centuries, but it will not always be so. Dubhe and Alkaid (Eta) are moving in different directions from the other stars in the Big Dipper. So Merak and Dubhe will end up pointing in quite a different direction.

Variables
Ursa Major has a number of variables that are binocular subjects. They include two long-period variables of the Mira-type: R, north of Dubhe; and T, which makes an equilateral triangle with Epsilon and Delta. They both vary from a respectable magnitude 6.7 down to beyond 13. R has a period of 301 days; T a period of 265 days.

Z, west of Delta, is a semi-regular variable, which always stays within binocular range, varying between magnitudes 7 and 9 in about 200 days.

A fine pair
Ursa Major is a region of the sky that is well-endowed with galaxies, many of them within reach of amateur telescopes and a few within reach of binoculars. The two brightest are a close pair, M81 and M82, which appear to be physically associated, or "interacting." They are perhaps best found from Dubhe and 23, with which they form a right-angled triangle. These two galaxies form part of a small cluster that lies relatively close to Earth, only about 8 million light-years away.

M81 is a classic spiral galaxy not dissimilar to our own, and bright at a magnitude of about 7. M82 is somewhat fainter and irregular in shape. It is a powerful radio source, which indicates that some almighty upheaval is occurring inside it.

Another galaxy of interest is the eighth-magnitude M101. It is a large spiral seen face-on, located north of the handle of the Big Dipper, and forms an equilateral triangle with Eta and Mizar. Large telescopes are needed to identify the spiral arms and to bring out nebulae therein. Among other galaxies relatively easy to find are K109, close to Gamma, and M108, close to Merak. Just south of M108 is a fine planetary nebula, M97, called the Owl.

For information about surrounding constellations, see the following pages:

Canes Venatici	50	Draco	40
Leo	48	Lynx	66

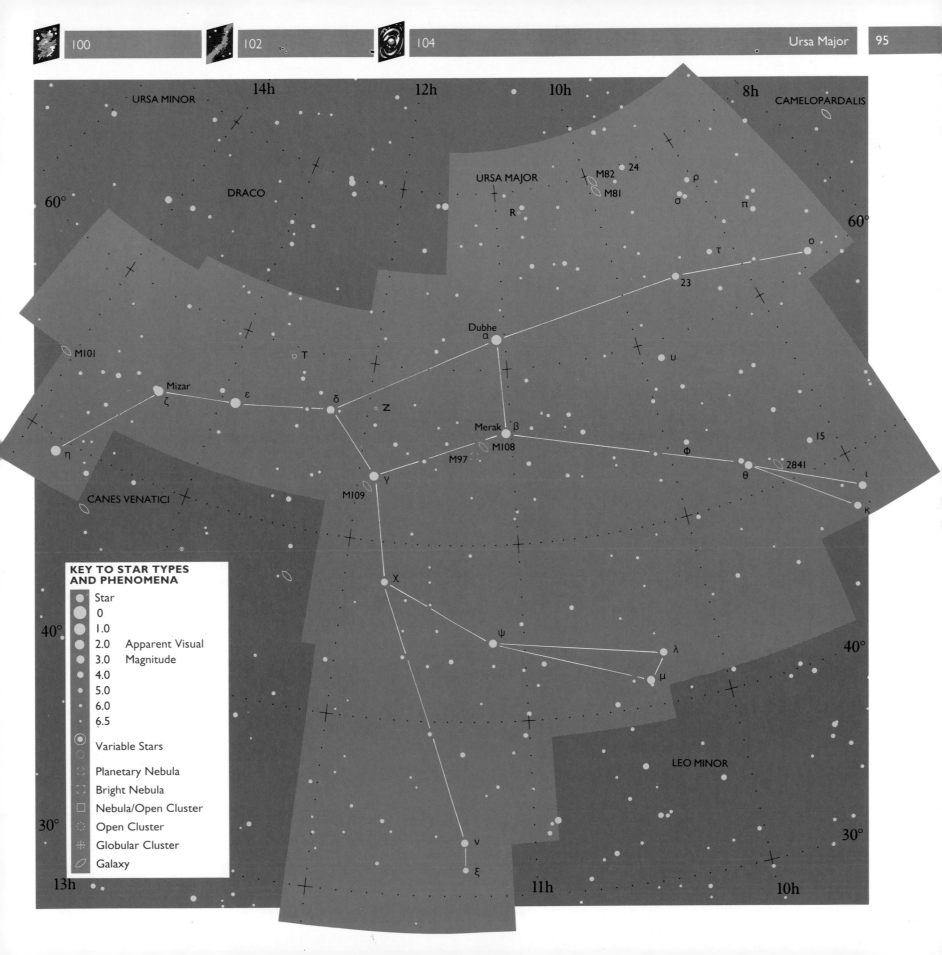

URSA MINOR

14h

12h

10h

8h

CAMELOPARDALIS

DRACO

60°

URSA MAJOR

M82　24

M81

ρ

R

σ

π

60°

τ

o

23

Dubhe
α

M101

u

T

Mizar

ζ　ε　δ

Z

15

η

Merak　β

φ

M108

M97

2841

θ

ι

CANES VENATICI

γ

κ

M109

**KEY TO STAR TYPES
AND PHENOMENA**

Star

0

1.0

2.0　Apparent Visual

3.0　Magnitude

4.0

5.0

6.0

6.5

Variable Stars

Planetary Nebula

Bright Nebula

Nebula/Open Cluster

Open Cluster

Globular Cluster

Galaxy

X

ψ

λ

40°

μ

40°

LEO MINOR

30°

ν

30°

ξ

13h

11h

10h

VIRGO

Virgo the Virgin represents the goddess of justice Astraea in Greek and Roman mythology. Earlier, she was the goddess Ishtar in Babylon and Isis in Egypt. Astraea is depicted with wings, with which she flew into the heavens when she despaired of the human race. In her left hand, she holds an ear of wheat marked by Spica.

Virgo spans the celestial equator and covers a vast region of the heavens. It is very nearly the largest constellation, second only to Hydra. It has a discernible Y-shape, but to see any resemblance to a maiden, with or without angel wings, requires considerable imagination.

To the naked eye, Virgo is disappointing, since it has only one really bright star. However, telescopes reveal the constellation to be crammed with deep-sky objects, particularly galaxies. They form part of a huge cluster that extends into Coma Berenices.

Virgo's brightest star is Alpha, or Spica, whose magnitude is exactly 1. It is bluish-white and is actually a binary, although this can only be detected from its spectrum.

Gamma and Epsilon are the next brightest stars, each of magnitude 2.8. Gamma is also a binary, with twin yellowish components (magnitude 3.6), which can be separated by small telescopes. The third-magnitude Delta, between Gamma and Epsilon, is noticeably reddish, as is Nu, north of Beta. Both Delta and Nu are red giants.

The Virgo Cluster

Extending right through the constellation, from south of Spica to the west of Epsilon, is a broad band of sky containing literally thousands of galaxies. They make up the celebrated Virgo Cluster.

This band of galaxies extends well into the neighboring constellation, Coma Berenices. The whole system, called the Coma-Virgo Cluster, contains at least 3,000 galaxies. It is one of the largest of the many groupings of galaxies that populate the Universe and also one of the nearest. The closest galaxies lie about 40 million light-years away, the farthest ones about 70 million light-years.

A number of the galaxies in the Virgo Cluster are bright enough to be visible in small telescopes, and a few may even be glimpsed with powerful binoculars. Most, however, require large telescopes for detection or for bringing out any structure.

The Messier galaxies

The French astronomer Charles Messier cataloged many of the brighter galaxies (averaging about the ninth magnitude) and these are worth pursuing. Several are to be found in the region between Epsilon Virginis and Beta Leonis. In order they are M60, M59, M58, M89, M90, M87, M84, and M86.

M60 and M59 are a close pair of elliptical galaxies due north of the fifth-magnitude Rho. M58, slightly to the west, is a barred-spiral galaxy. The elliptical M89 and spiral M90 lie to the north of M58, and M87 lies to the west. The close pair M84 and M86, both ellipticals, appear in the same telescopic field of view north of M87.

Among these galaxies, M87 is outstanding in more ways than one. It is a giant elliptical galaxy, which is nearly spherical; its diameter is similar to that of our own galaxy, about 120,000 light-years, but our galaxy is essentially a disk, so the near-spherical M87 contains very many more stars. Indeed, it is one of the most massive galaxies known and is a powerful source of radio waves.

Another interesting Messier galaxy, M104, an edge-on spiral, is found right on the border with Corvus, due south of Chi. It is of the eighth magnitude and may just be glimpsed in binoculars. Larger telescopes or long-exposure photographs are needed to bring out the dark dust lane that cuts through the oval image, giving it the vague appearance of a hat. This is why it is known as the Sombrero Galaxy.

Quasar quest

Beyond the range of small telescopes, but detectable with large ones, is an object called 3C373, located at the corner of a right-angled triangle with Gamma and Eta Virginis. It is a quasar, and is of great historical interest and importance.

3C273 was the first quasar to be recognized as such, in 1962. A very remote star-like object with an incredible energy output, it shines at about the thirteenth magnitude, yet it is over 3,000 million light-years away. No ordinary galaxies would show up at such a distance.

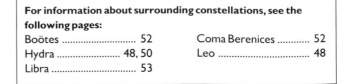

▲ The galaxy M87, which may perhaps contain as many as 3 million million stars, 30 times as many as our own Galaxy. This near-spherical object is one of the most notable "active" galaxies, pouring out enormous energy, particularly at radio and X-ray wavelengths.

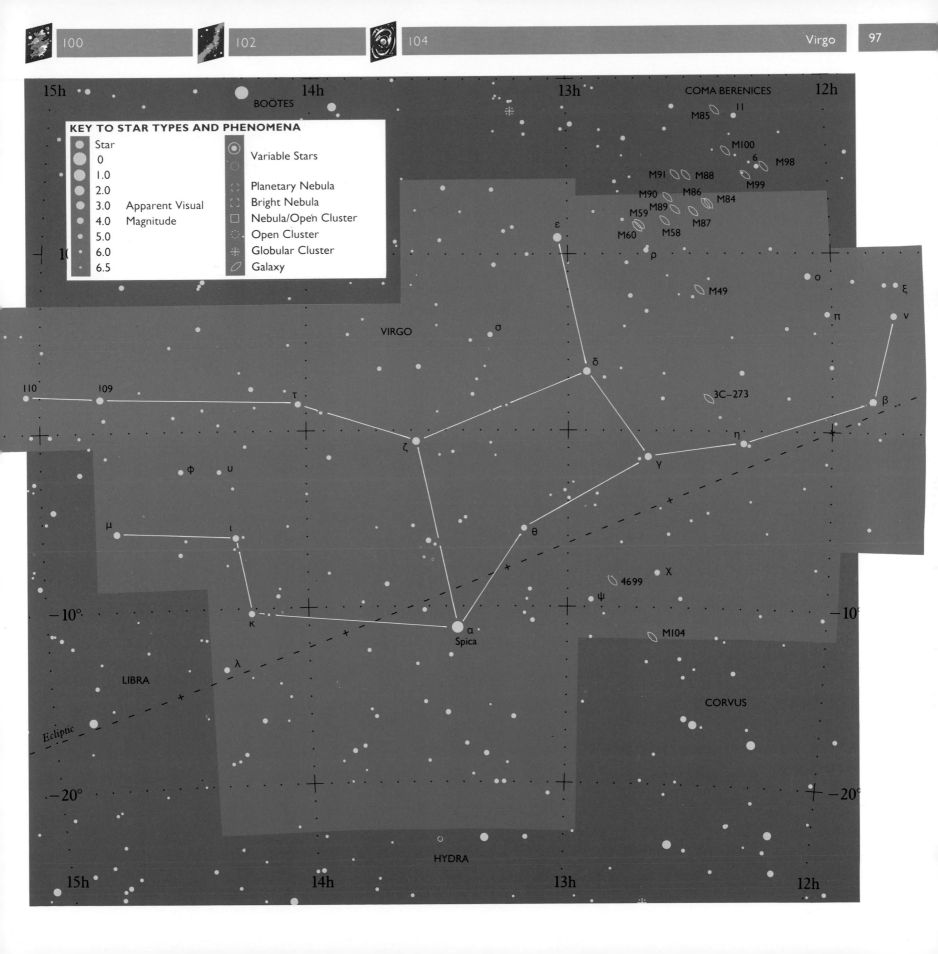

15h   14h   13h   COMA BERENICES   12h

BOOTES

KEY TO STAR TYPES AND PHENOMENA

Star
- 0
- 1.0
- 2.0
- 3.0   Apparent Visual
- 4.0   Magnitude
- 5.0
- 6.0
- 6.5

⊙ Variable Stars
◇ Planetary Nebula
▢ Bright Nebula
□ Nebula/Open Cluster
⊙ Open Cluster
✳ Globular Cluster
⬭ Galaxy

M85   11
M100
ô   M98
M91   M88
M90   M86   M99
M59   M89   M84
M60   M58   M87

ρ

o

ξ

ε

VIRGO

σ

M49

δ

π

ν

110   109   τ

3C–273

β

ζ

η

φ   υ

γ

μ   ι

θ

4699   X

ψ

−10°   κ   α   M104   −10°

Spica

λ

LIBRA

CORVUS

−20°   −20°

Ecliptic

HYDRA

STAR CLUSTERS

As we observed in an earlier chapter, stars are gregarious and tend to travel through space with one or more companions. Sometimes they form part of much larger groupings of stars which we call clusters.

The most prominent cluster in the Northern Hemisphere is the Pleiades, or Seven Sisters, in Taurus. In the Southern Hemisphere, the most prominent cluster is Omega Centauri, in Centaurus. But these two clusters are fundamentally different. The Pleiades is a fairly loose grouping of stars known as an open, or galactic cluster. Omega Centauri is a densely packed "globe" of stars known as a globular cluster.

Open clusters

We can see up to seven stars in the Pleiades with the naked eye (hence its popular name), but binoculars or a small telescope will reveal that it consists of many more stars. Altogether it contains several hundred, and this number is typical of an open cluster.

Although the stars in the Pleiades appear to be close together, they are in fact quite far apart. The whole cluster occupies a region about 15 light-years across.

▼ The Pleiades, or Seven Sisters, one of the two striking clusters in Taurus. It is easily seen with the naked eye and looks glorious in binoculars. Its brightest star, Alcyone (in the center of the picture), is just above the third magnitude.

▲ This fine open cluster in Crux is NGC4755, popularly called the Jewel Box. It is often named the Kappa Crucis cluster after its central, bright red star.

Long-exposure photographs of the Pleiades reveal that the bright stars are surrounded by a diffuse gas cloud, or nebulosity. This is typical of open clusters, which are usually made up of young, hot stars.

Taurus is also the home of another easily seen open cluster, the Hyades. This conglomeration of stars forms a V-shape around the brilliant orange Aldebaran, although this star is not part of the cluster – it is very much closer to us. The Hyades itself is one of the closest open clusters. About 140 light-years away, it is only a quarter of the distance to the Pleiades.

Several other open clusters make rewarding viewing, among them M44 in Cancer. It is named Praesepe, or the Beehive, because it has been likened to a swarm of bees buzzing around a hive. Perseus boasts a close pair of open clusters, the famous Double Cluster (right), which looks magnificent in binoculars. In far southern skies is the incomparable Jewel Box in Crux (above), which lies close to the dark nebula we call the Coal Sack. Another southern cluster is M11 in Scutum, whose wing-shaped arrangement has earned it the name of the Wild Duck.

Globular clusters

The globular cluster Omega Centauri (page 82) could not be more different from the Pleiades open cluster. It consists of literally hundreds of thousands of stars packed together in a region of space only about 60 light-years across. To the eye, it looks like one bright star, which is

◄ The globular cluster M3 in Canes Venatici. One of the top globulars in northern skies, it is just beyond naked-eye range, but an easy binocular subject. Small telescopes will begin to resolve the outer fringes of this globe of some 45,000 stars. Like most globulars, M3 is remote, lying over 40,000 light-years away.

▼ The distinctive Double Cluster in Perseus, also called the Sword Handle. These two clusters are naked-eye objects, designated h (left) and Chi Persei.

why it was given a star name. But binoculars will start to reveal individual stars in the outer regions, and telescopes will resolve a host more.

Omega Centauri differs from the Pleiades in many other respects. For example, it lies 40 times farther away, at a distance of about 16,000 light-years. It is made up of very old, not young, stars. And, whereas the Pleiades rotates with the other stars in the plane of our Galaxy, Omega Centauri pursues quite a different orbit that takes it above and below the galactic plane.

Omega Centauri is typical of globular clusters. They are all dense globes of very old stars – some appear to be nearly as old as the Galaxy itself. They are all remote and circle in quite different orbits from ordinary stars within the galactic "halo" – a spherical region of space around the center of the Galaxy.

A few of the other 150 or so globular clusters are visible to the naked eye. Rivalling Omega Centauri in magnificence in the Southern Hemisphere is 47 Tucanae in Tucana. In the Northern Hemisphere, M13 in Hercules is outstanding, though only just bright enough to be visible to the naked eye. M92, also in Hercules, M3 in Canes Venatici (above), and M15 in Pegasus are somewhat fainter, but are excellent subjects for small telescopes.

NEBULAE

The universe, as we see it, consists of countless galaxies, separated from one another by vast expanses of empty space. In turn, the galaxies consist of billions of stars, separated from one another by apparently empty space.

Space, however, is not completely empty. It contains scattered traces of matter in the form of molecules of gas and particles of dust. In general, this interstellar matter is thinly distributed and difficult to detect, but in places it clumps together into denser clouds, which astronomers call nebulae (after the Latin word for cloud, *nebula*). Even such "dense" clouds, however, are a million million million times less dense than air!

Bright nebulae

We can see many nebulae in binoculars and telescopes, and a few even with the naked eye, glowing brightly against the dark backdrop of space. The easiest to spot is the Great Nebula in Orion (below), which is visible to skywatchers in both Northern and Southern Hemispheres.

This billowing, flaming cloud is an example of an emission nebula. In this type, gas particles in the cloud give out light when they are excited by (absorb energy from) radiation from hot stars embedded within them.

Like most emission nebulae, the Orion Nebula is tinged

red, and there is a very good reason for this. Red is the color of the light emitted by hydrogen atoms when they become excited. And hydrogen is the most plentiful element in nebulae.

Other emission nebulae are also spectacular, though most require larger telescopes and long-exposure photography to bring out their true splendor. Many have distinctive shapes that have earned them their names. For example, the Trifid Nebula in Sagittarius has three dark dust lanes running through it; the North America Nebula in Cygnus (right) could not be better named.

When the stars within or near a nebula are cool, their radiation is not energetic enough to excite its atoms. Then the nebula merely reflects the stars' light. Such a reflection nebula is thus relatively faint, and can usually be seen as a diffuse patch of light. The young hot stars in the Pleiades star cluster in Taurus (see page 99) are surrounded by nebulosity of this type.

▼ The Great Nebula (M42) in Orion is perhaps the finest nebula of all, and close enough to be seen with the naked eye. The smaller separate nebula above the Great Nebula is named M43, but it is really an extension of the same nebulosity.

▲ A dramatic close-up of the most famous dark nebula, the aptly named Horsehead Nebula, in Orion. The mass of dark matter is silhouetted against the bright nebula IC434, which extends south from Zeta Orionis, in Orion's Belt.

Dark nebulae

There are many clouds of gas and dust in the heavens that are not lit up by nearby or embedded stars. But we can still sometimes detect them, when they blot out the light from more distant stars or nebulae. To our eyes, they appear black, and are termed dark nebulae. Best-known are the Horsehead Nebula in Orion (opposite) and the Coal Sack in Crux.

Astronomers reckon that invisible dark matter makes up quite a large proportion of the mass of the Universe and that the amount of dark matter there is will determine the future of the Universe.

Death throes

Other types of nebulae result when stars approach the end of their lives. A star that has expanded into a red giant often blows off gas, which travels out through space as an expanding spherical shell or bubble. Through a telescope, the gas bubble appears as a diffuse disk, rather like a planet. That is why we call this type a planetary nebula. The Ring Nebula in Lyra (below), the Helix Nebula in Aquarius, and the Dumbbell Nebula in Vulpecula are beautiful examples of the genre.

A writhing cloud of expanding gas also results from the spectacular death of a supergiant star in a supernova (see page 73). The classic example is the Crab Nebula in Taurus (see page 92), the remains of a supernova witnessed by Chinese astronomers in the year 1054. The Veil Nebula in Cygnus (see page 84) is another beautiful example, aptly named after its pattern of glowing filaments, which resemble delicate lacework.

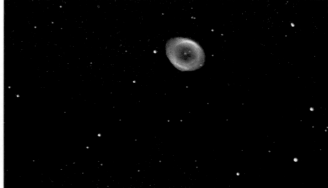

◀ This chaotic area of stars and glowing gas bears an uncanny resemblance to the continent after which it is named, North America. The nebula (NGC7000) is in Cygnus, and can be spotted in binoculars close to Deneb.

▲ The Ring Nebula (M57) in Lyra is a colorful "smoke-ring" of glowing gas and is one of the finest planetary nebulae in the heavens. The ring is made up of gas puffed out by the central star, probably about 6,000 years ago.

THE MILKY WAY

On dark, clear nights, when there is no Moon, you can see a hazy band of light spanning the dome of the heavens. This is aptly named the Milky Way.

The Milky Way passes through some of the most brilliant constellations: Cassiopeia, Cygnus, Perseus, Auriga, and Aquila in the Northern Hemisphere, and Puppis, Vela, Carina, Crux, Centaurus, Scorpius, and Sagittarius in the Southern Hemisphere. It varies noticeably in brightness. In the Northern Hemisphere, it is brightest in Cygnus and Aquila; in the Southern Hemisphere, in Scorpius and Sagittarius. The Milky Way also varies in width, in places being only about 5° across, while in others approaching 30°.

But what is the Milky Way? No one knew until Galileo turned his telescope on this heavenly band nearly 400 years ago. He was amazed at what he saw, as we are today when we look at it in binoculars. The milky white band resolves into a mass of faint stars beyond number.

In places, for example in Cygnus and Aquila, the Milky Way is split by dark rifts where there appear to be scarcely any stars. The stars are there, but intervening clouds of dust are blotting out their light.

The Milky Way represents a cross section of the great star system, or galaxy, to which our Sun and all the other stars in the sky belong. We often refer to this system as the Milky Way, but usually we just call it "the Galaxy."

▼ A map of the whole night sky, based on information sent back by the astronomy satellite IRAS, which pictured the Universe in infrared light. It shows, in effect, a cross section of our Galaxy: the bright region across the middle is the star-studded Milky Way. The large and small white patches underneath are our closest galactic neighbors, the Large and Small Magellanic Clouds.

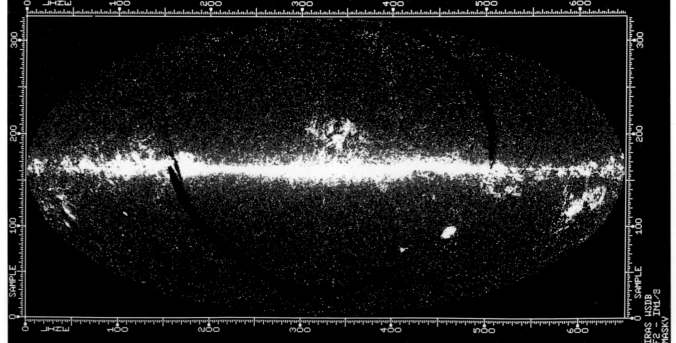

▲ A view of the Milky Way in the southern constellation Norma, showing dark rifts where dust is obscuring the light from stars behind. This picture, taken in March 1986, shows a celestial interloper, Halley's Comet, which was seen best in the Southern Hemisphere that year.

▶ Plan and side views of our Galaxy, a typical and quite large spiral. Although most matter is concentrated in the spiral disk, traces exist in a spherical "halo" that surrounds it.

◄ Justly the most famous of all the outer galaxies, the Great Spiral in Andromeda, M31. Visible to the naked eye, its spiral structure mirrors that of our own Galaxy.

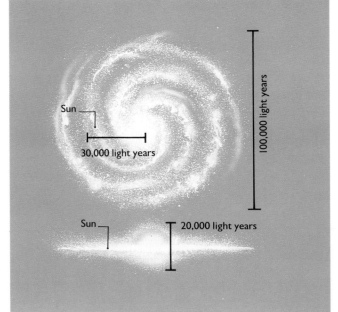

Sun

30,000 light years

100,000 light years

Sun   20,000 light years

It takes the form of a disk with a bulge at the center (the nucleus). From the nucleus, a number of arms spiral out, carrying the stars. The whole Galaxy rotates and, viewed from afar, would perhaps present the appearance of a Catherine wheel firework (see left).

The size of our Galaxy is astounding. It measures 100,000 light-years across and is made up of at least 100,000 million stars. But it is nothing special among galaxies: it is a typical spiral galaxy of the Sb type (see page 104). Many galaxies, including the Great Spiral in Andromeda (above), are much bigger.

The Sun is located on one of the spiral arms about 30,000 light-years from the center of the Galaxy. Like the other stars, it rotates around the galactic center, completing one orbit in 225 million years, a period called the cosmic year.

The center of the Galaxy lies in the direction of Sagittarius, and that is why the Milky Way is densest there and in the neighboring constellations. We cannot see into the center because of obscuring dust clouds.

▲ Seen edge-on, our own Galaxy would probably look much like this one. It is NGC5907 in Draco, a spiral of the Sb type. Note how the disk thickens toward the center.

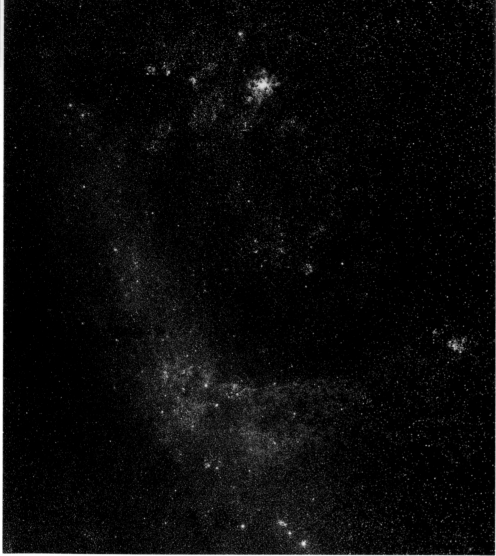

▲ This galaxy is NGC5128 in Centaurus, also called Centaurus A. It is a bright spherical galaxy only just out of naked-eye range. It is notable for the dark dust lane that divides it in two.

▼ Centaurus A in a different light: as it appears at radio wavelengths. It is the third most powerful radio source in the heavens, a sure sign that something devastating is happening inside.

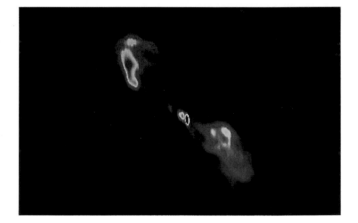

▲ The Large Magellanic Cloud in Dorado, one of the great delights of the Southern Hemisphere. It is visible as a misty patch to the naked eye, but easily resolved in telescopes into individual stars. Near the top of this fine photograph is the Tarantula Nebula, one of the largest nebulae we know.

THE OUTER GALAXIES

In the 18th century, the great "collector" of nebulous objects, Charles Messier, listed as number 31 in his catalogue a fuzzy patch visible to the naked eye north of the star Beta Andromedae. It was called the Great Nebula in Andromeda (see page 103).

Only later, when telescopes became more powerful, was it revealed that this "nebula" was actually made up of stars, star clusters, and nebulae. It was no nebula of our own Galaxy, but an entirely separate galaxy. When its distance was measured, this was confirmed. The "nebula" lies over two million light-years away, very much farther than any of the stars in our Galaxy.

Two other naked-eye "nebulae" also prove to be separate galaxies. They are the Large Magellanic Cloud (LMC) in Dorado and the Small Magellanic Cloud (SMC) in Tucana, which lie close to the southern celestial pole. The LMC, also called Nubecula Major, is in fact the closest galaxy to our own, being only about 170,000 light-years away. As far as distances in the Universe are concerned, this is a stone's throw.

Compared with the Great Spiral in Andromeda, the LMC is tiny. The Andromeda Galaxy measures more than 150,000 light-years across, half as big again as our own Galaxy and more than five times the size of the LMC.

Classifying the galaxies
The Andromeda Galaxy and the LMC are also poles apart in structure. The first has a fairly regular spiral form,

while the other is quite irregular and shows little structure. These two galaxies typify two major classes of galaxies, the spirals and the irregulars. The third major class of galaxies is elliptical, which covers spherical as well as oval-shaped galaxies.

This classification of galaxies by their shape was suggested by the great pioneer galaxy observer Edwin Hubble in the 1920s. Hubble graded ellipticals E0 to E7 according to how oval they were. He graded spiral galaxies (S) according to the openness of their spiral arms: Sa, Sb, and Sc, Sc indicating wide-open arms. Our own Galaxy and Andromeda are both classed as Sb. M33 in Triangulum (page 65) is classed as Sc. Some spiral galaxies have a kind of bar passing through the nucleus and are classed as barred spirals.

Active galaxies

The majority of galaxies seem to be "normal," giving off the light and energy we would expect for a grouping of billions of stars. But some galaxies have an extraordinary energy output resulting from some extraordinary process going on inside. We call them active galaxies.

Many of the active galaxies send out most of their energy at radio wavelengths. An example is NGC5128, or Centaurus A (left), one of the first radio galaxies to be discovered. Strangely, the energy from a radio galaxy appears to be radiated from a region of space far removed from the galaxy itself!

Clusters of galaxies

Galaxies as well as stars tend to cluster together in space. Our own Galaxy forms part of a cluster called the Local Group. Its 30 or so members also include the Large and Small Magellanic Clouds, the Andromeda Galaxy, and M33 in Triangulum.

The Local Group occupies a region of space about 5 million light-years across. It is a relatively small cluster. The next cluster to us contains up to 3,000 galaxies! It is the Coma-Virgo Cluster, named after the constellations in which it is found (Coma Berenices and Virgo). It is centered on the bright active galaxy M87.

But the clustering does not stop there. Our Local Group and the Coma-Virgo Cluster are just part of a huge supercluster of galaxies. There are many such superclusters in the Universe.

◀ Some of the more prominent members of the Coma-Virgo Cluster, which contains galaxies by the thousand. The largest ones in the picture are a closely matched pair of elliptical galaxies M86 (top) and M84, which are located right at the center of the Cluster. They are both of the ninth magnitude.

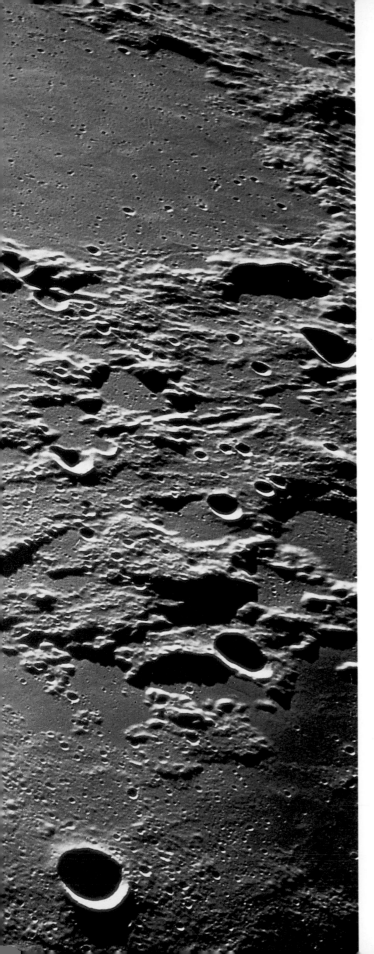

◀ A stunning landscape close to the center of the Moon, pictured by Apollo astronauts in 1969. The large crater at top right is Ptolemaeus, more than 90 miles (145 kilometers) across. The small crater on its left is Herschel. The similar-sized crater near the bottom of the picture is Lalande.

THE MOON

The distant stars and nebulae and even more remote galaxies are a source of constant delight for the astronomer. But there are plenty of astronomical delights much nearer home. Right on our celestial doorstep is the Moon, the Earth's only natural satellite and by far its closest neighbor in space.

The photograph hints at the pleasures that await us when we turn our binoculars and telescopes on our silvery satellite: craters, flat plains, rugged highlands, wrinkle ridges, and still more craters – always craters.

Being our satellite, the Moon circles around the Earth, taking a little under a month to do so. During this time, it appears to change shape, revealing first more and then less of its surface as it waxes and wanes. The greatest detail can be seen on the terminator, the boundary between light and dark.

The maps in this section show the surface of the Moon in four quadrants. We show it the "right way up," as it appears to the naked eye and through binoculars. Through a telescope, however, these views of the Moon will appear upside-down.

▲ A gibbous Moon, snapped by Apollo astronauts as they sped toward it. The photograph shows well the complex of seas in the eastern hemisphere: Mare Crisium, Serenitatis, Tranquillitatis, Fecunditatis, and Nectaris. Use the Moon maps that follow to identify the bright craters.

▶ The Earth rises above the lunar horizon in one of the classic space photographs taken by the Apollo astronauts. How colorful and inviting the Earth looks compared with the rugged and crater-strewn landscape below.

QUEEN OF THE NIGHT

The Moon is the Earth's constant companion as it journeys through space – its only natural satellite. Although it is relatively small, the Moon dominates the night sky because it is so much closer than any other heavenly body. But the Moon has no light of its own. It shines because it reflects sunlight.

The Moon lies on average about 239,000 miles (384,000 kilometers) away from the Earth and takes about a month to circle around it. As the Moon circles, we see it go through its phases, with more or less of its surface lit up. The main phases are New Moon, First Quarter, Full Moon, and Last Quarter. Between New Moon and First Quarter, and between Last Quarter and New, the Moon appears as a crescent. Between First Quarter and Full, and between Full and Last Quarter, its shape is described as gibbous.

With a diameter of 2,160 miles (3,476 kilometers), the Moon is about a quarter the size of the Earth. This is extraordinarily large for a satellite in relation to its parent planet. So astronomers sometimes consider Earth-Moon as a double planetary system.

Being so small, the Moon has, compared with the Earth,

a low mass and therefore weak gravity (one-sixth as strong as the Earth's). As a result, it does not have enough "pull" to hold onto an atmosphere. The airless Moon has no water either. Without an atmosphere, the Moon also experiences a marked difference in temperature between its two-week-long days and nights. In the lunar day temperatures soar to over 212°F (100°C), while in the lunar night they plummet below –240°F (–150°C).

Observing the Moon
Because the Moon appears so large in the night sky, it is the ideal target for the astronomer, and it is popular with newcomer and old hand alike. The simplest of equipment, binoculars or a small telescope, will reveal a wealth of detail.

Even the naked eye reveals two distinct regions on the Moon – dark and light. The dark ones are the maria, or seas. The light ones are rugged highland areas. Again, the naked eye reveals that the Moon always presents the same face to us: the near side. We can never see the far side from Earth. Any optical aid will show the Moon's other major features, its craters and its mountains.

The fascinating thing about lunar observing is that the appearance of the Moon changes nightly as it goes through its phases. In particular, the boundary between light and shadow – lunar day and night – moves over the surface. We call this boundary the terminator. It sweeps over the lunar surface twice – from New to Full, and from Full

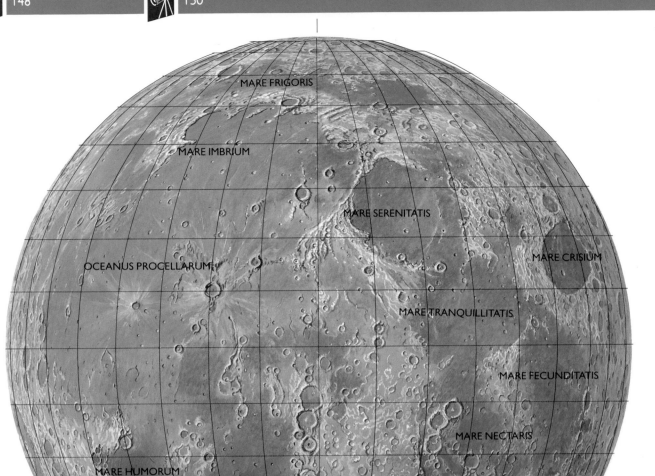

MARE FRIGORIS

MARE IMBRIUM

MARE SERENITATIS

OCEANUS PROCELLARUM

MARE CRISIUM

MARE TRANQUILLITATIS

MARE FECUNDITATIS

MARE NECTARIS

MARE HUMORUM

MARE NUBIUM

MARE MOSCOVIENSE

Mendeleev

Gagarin

Tsiolkovsky

Hertzsprung

Korolev

Leibnitz

Apollo

◄ The near side of the Moon, which always faces us. It is shown the "right way up", that is, as we see it with the eye and in binoculars. The greater part of the surface is covered with "seas," or maria, which appear darker than the highly cratered highlands. The highlands are thought to be part of the Moon's ancient crust. The maria are younger, formed when gigantic meteorites slammed into the surface and caused the rocks to melt again.

▼ The far side of the Moon, which we can never see from Earth. It remained an enigma until space probes began photographing it, beginning with the Russian Luna 3 probe in 1959. The strange thing about the far side is that it is devoid of any large maria. One of its most recognizable features is a large dark-floored crater named Tsiolkovsky.

back to New, as the Moon waxes and wanes.

It is on the terminator that the greatest detail can be seen. Because of the low Sun angle, long shadows emphasize the craters, mountains, and other surface features dramatically. By contrast, Full Moon is not a good time for detailed observation, because the Sun is high up and casts scarcely any defining shadows.

The Moon maps that follow cover the lunar surface in four quadrants. North is at the top. The maps show the most outstanding features of the lunar landscape. The names of the maria and like features are given in Latin. The English equivalents can be found in the Glossary at the end of the book.

Northwest Quadrant

Oceanus Procellarum and Mare Imbrium dominate this quadrant. Measuring more than 700 miles (1,100 kilometers) across, Mare Imbrium is the largest of the circular seas. Its boundaries are well defined and marked by high mountain ranges – the Carpathians, Apennines, Caucasus, Alps, and Jura mountains.

The Carpathians, which peak at about 7,000 feet (2,100 meters) extend northwest from Copernicus and form the boundary between Mare Imbrium and Oceanus Procellarum. The Apennines extend for some 280 miles (450 kilometers) and contain some of the highest peaks on the Moon, which soar in places to more than 20,000 feet (6,000 meters). The Caucasus Mountains separate Mare Imbrium from Mare Serenitatis to the east.

Sinus Iridum is a spectacular bay of Mare Imbrium, measuring some 150 miles (250 kilometers) across. It is a spectacular sight on the terminator when the Moon is about 10 days old.

Outstanding craters and other features
Archimedes (47 miles, 75 kilometers) is the largest of a prominent trio of craters, the others being Autolycus and Aristillus. All have flat, lava-filled floors.
Aristarchus is only 23 miles (37 kilometers) across, but

is the brightest feature on the Moon. Nearby is a deep winding fault called Schroter's Valley.
Copernicus (60 miles, 97 kilometers) is one of the most conspicuous lunar craters. It has a classic profile: high terraced walls and central mountain peaks. At Full Moon it is surrounded by bright crater rays.
Eratosthenes is a two-thirds size replica of Copernicus and lies at the end of the Apennine chain.
Kepler is small (22 miles, 35 kilometers) but stands out prominently on Oceanus Procellarum at Full Moon because of its sparkling crater rays.
Lansberg (26 miles, 42 kilometers), on the lunar equator, stands out clearly on Oceanus Procellarum. It is a near-twin of Reinhold to the northeast.
Otto Struve (100 miles, 160 kilometers) is one of the largest craters in this quadrant, but it is difficult to see because it lies close to the western limb (edge).
Pico is an isolated mountain near the edge of Mare Imbrium, just south of Plato. It rises to about 8,000 feet (2,400 meters).
Plato (60 miles, 97 kilometers), on the edge of the Alps, is a circular crater, noted for its flat, very dark floor.
Reinhold (30 miles, 48 kilometers) lies between Copernicus and Lansberg. Like the latter, it has a deep floor.
Straight Range is an isolated mountain range near the northern edge of Mare Imbrium, close to Plato. It measures about 40 miles (60 kilometers) long and has peaks up to about 6,000 feet (1,800 meters).

▲ Two of the quadrant's most prominent craters, both of which have a characteristic central mountain peak. They are Copernicus (top) and Eratosthenes. The terraced walls of Copernicus are clearly evident.

► Another of the quadrant's spectacles, the crater Kepler, as pictured by the Apollo 12 astronauts from lunar orbit. Note how smooth the surrounding mare surface is.

Most of the northwest quadrant is occupied by three seas, the sprawling Oceanus Procellarum, Mare Imbrium, and in the north, Mare Frigoris. Oceanus Procellarum is the largest sea on the Moon. Unlike most of the others, it has no definite boundary. Among the craters that grace this quadrant, Copernicus, Kepler, and Aristarchus are outstanding, being at the hub of spectacular ray systems at Full Moon.

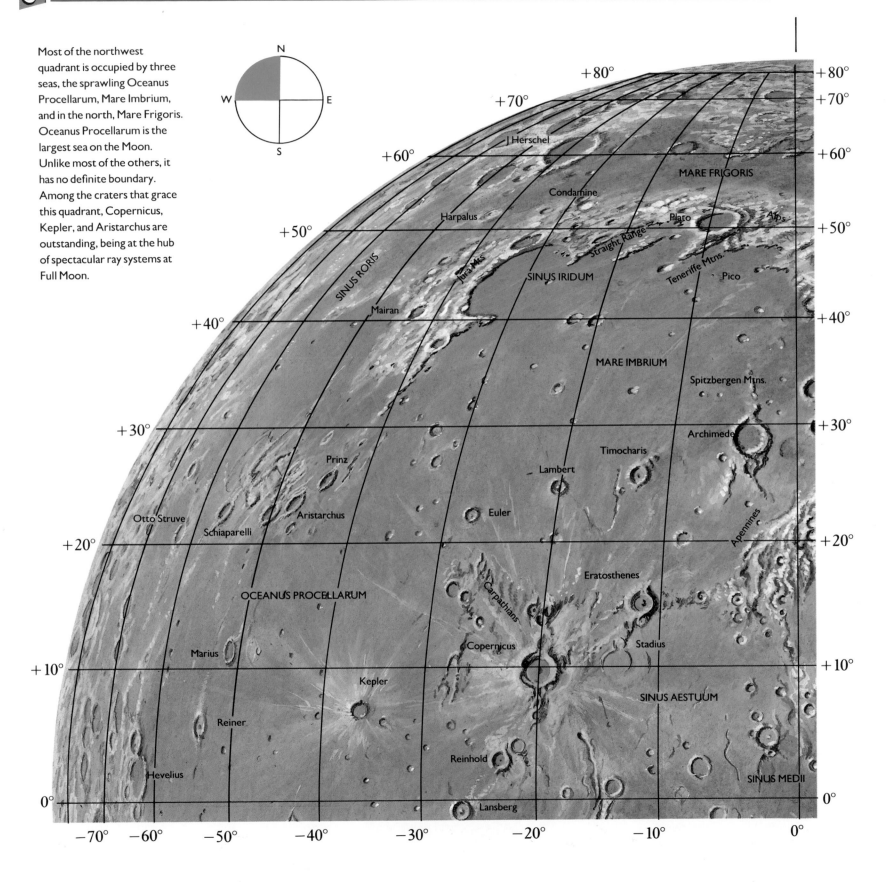

N
W   E
S

+80°
+70°
+60°
+50°
+40°
+30°
+20°
+10°
0°

J Herschel
MARE FRIGORIS
Condamine
Harpalus
Plato
Alps
Straight Range
SINUS RORIS
Jura Mts
Teneriffe Mtns.
Pico
SINUS IRIDUM
Mairan
MARE IMBRIUM
Spitzbergen Mtns.
Archimedes
Prinz
Timocharis
Lambert
Otto Struve
Euler
Schiaparelli
Aristarchus
Apennines
OCEANUS PROCELLARUM
Eratosthenes
Carpathians
Marius
Copernicus
Stadius
Kepler
SINUS AESTUUM
Reiner
Reinhold
Hevelius
SINUS MEDII
Lansberg

−70°   −60°   −50°   −40°   −30°   −20°   −10°   0°

Southwest Quadrant

Mare Cognitum is often considered to be a bay of Oceanus Procellarum, a sea that covers an area of some 2 million square miles (5 million square kilometers). This is more than half as big again as the Mediterranean Sea on Earth.

Mare Humorum is one of the smaller circular seas, measuring about 280 miles (450 kilometers) across. It is seen best when the Moon is about 11 days old. Mare Nubium is less well defined, but boasts a more interesting linear feature called the Straight Wall. Running for about 60 miles (100 kilometers), it marks a fault in the Moon's crust. The Wall is about 800 feet (240 meters) high.

Two of the six Apollo landings were made in this quadrant a few degrees south of the lunar equator:

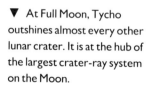

► A close-up picture of the splendid southern crater Tycho, taken by a US Lunar Orbiter probe from a distance of about 135 miles (220 kilometers). Note the crater's terraced walls and the central mountain peak.

▼ At Full Moon, Tycho outshines almost every other lunar crater. It is at the hub of the largest crater-ray system on the Moon.

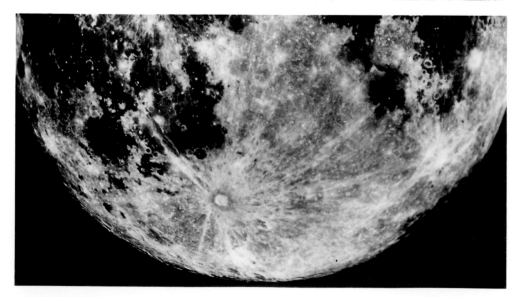

Apollo 12 set down on Oceanus Procellarum and Apollo 14 near the Fra Mauro formation.

Outstanding craters and other features

Alphonsus (80 miles, 129 kilometers) is the middle of three large walled plains near the 0° longitude line. Rilles and mountains cross its floor. It is best seen at First Quarter.

Bailly is the largest crater on the Moon, with a diameter of 183 miles (295 kilometers). Unfortunately, it is right on the limb (edge) and is poorly seen at any time.

Bullialdus (31 miles, 50 kilometers), on the edge of Mare Nubium, is a perfectly formed crater with intact walls and central peaks.

Clavius is the largest lunar crater that we can see well from Earth, with a diameter of 144 miles (232 kilometers). Its walls and floor are peppered with craters.

Gassendi (55 miles, 89 kilometers) bordering Mare Humorum, is one of the many fine walled plains this quadrant has to offer. Its floor shows interesting detail.

Grimaldi (120 miles, 193 kilometers) lies a few degrees south of the equator, close to the western limb. It has low walls, but it is always easy to spot because of its dark floor. Indeed, it is the darkest area of the Moon.

Longomontanus (90 miles, 145 kilometers) is one of the trio of large craters in the south, the others being Clavius and Maginus. Like them, it has partly ruined walls.

Maginus (110 miles, 177 kilometers) lies due east of Longomontanus. These two craters, together with Clavius to the south and Tycho to the north, make a kind of lunar "southern cross."

Pitatus (50 miles, 80 kilometers), at the southern edge of Mare Nubium, has a dark floor with a low central peak.

Ptolemaeus is the first and largest of the string of large craters extending north–south near the 0° longitude line, which are particularly prominent at First Quarter. Close to the Moon's center, it is a walled plain with a dark floor measuring 92 miles (148 kilometers) across.

Riccioli (100 miles, 160 kilometers) lies just south of the equator near the western limb. Like its neighbor and near-twin Grimaldi, it has an exceptionally dark floor.

Schickard (124 miles, 200 kilometers) is a fine walled plain, with low surrounding walls.

Tycho vies with Copernicus for being the finest lunar crater. Though not big (52 miles, 84 kilometers), it becomes truly spectacular at Full Moon as the center of a bright crater-ray system, whose shining spokes extend in all directions, as far as Oceanus Procellarum to the northwest and Mare Nectaris to the northeast.

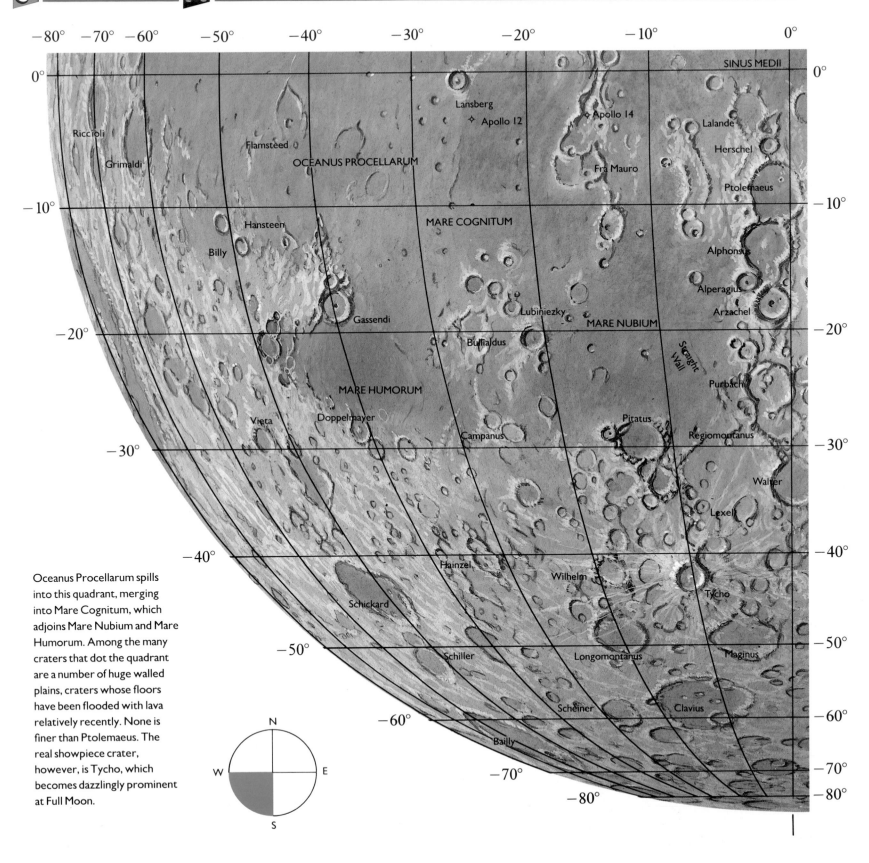

−80° −70° −60° −50° −40° −30° −20° −10° 0°

SINUS MEDII

0°   0°

Lansberg
⬦ Apollo 12   ⬦ Apollo 14   Lalande

Riccioli   Herschel

Flamsteed   Fra Mauro
Grimaldi   OCEANUS PROCELLARUM   Ptolemaeus

−10°   −10°
                  Hansteen   MARE COGNITUM
                                                        Alphonsus
Billy
                                                        Alperagius
                  Gassendi   Lubiniezky   Arzachel
−20°   MARE NUBIUM   −20°
                              Bullialdus
                                                        Straight Wall
                  MARE HUMORUM   Purbach
Vieta   Doppelmayer   Pitatus
                                  Campanus   Regiomontanus
−30°   −30°
                                                        Walter

                                                        Lexell
Oceanus Procellarum spills
into this quadrant, merging
−40°   −40°
into Mare Cognitum, which   Hainzel   Wilhelm
adjoins Mare Nubium and Mare   Tycho
Humorum. Among the many
craters that dot the quadrant
are a number of huge walled   Schickard
plains, craters whose floors
have been flooded with lava   Schiller   Longomontanus   Maginus
−50°   −50°
relatively recently. None is
finer than Ptolemaeus. The
real showpiece crater,   Scheiner   Clavius
                      N   −60°
however, is Tycho, which   −60°
becomes dazzlingly prominent
at Full Moon.   W ⊕ E   Bailly   −70°
                                                          −70°
                      S   −80°
                                                          −80°

Northeast Quadrant

The circular Mare Serenitatis measures about 435 by 560 miles (700 by 900 kilometers). It is bordered in the northwest by the Caucasus Mountains, which contain peaks of over 20,000 feet (6,000 meters).

The Haemus Mountains to the south are a much lower range. They give way in the east to an area of low hills and small craters, of which the largest is Plinius. The surface of the mare is noticeably wrinkled, but features only one prominent crater, Bessel.

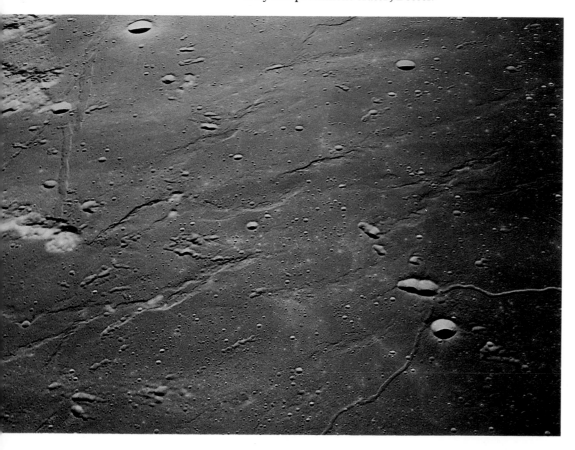

▲ A breathtaking view of the southwest region of Mare Tranquillitatis, close to the site of the first Apollo landing on the Moon. The Apollo 11 lunar module set down near the top center of the photograph. Maskelyne Rille is in the foreground.

In the north, Mare Serenitatis merges, through Lacus Mortis and Lacus Somniorum, into Mare Frigoris, which extends west into Oceanus Procellarum in the adjacent quadrant. The hilly region in the south of Serenitatis forms an indistinct border between that sea and the similar-sized Mare Tranquillitatis. This sea in turn merges into the also irregular Mare Fecunditatis near the equator.

The most distinctly circular sea in this quadrant, however, is Mare Crisium, near the eastern limb. Measuring about 370 by 290 miles (590 by 460 kilometers) across, it is marked by a few small craters, the largest being Picard.

At about the same latitude, but on the other side of the quadrant, is the small sea Mare Vaporum, which is about 280 miles (450 kilometers) across. It extends south toward the small mare region at the Moon's center, Sinus Medii.

Three Apollo landings took place in this quadrant: Apollo 11 landed near the southwestern edge of Mare Tranquillitatis; Apollo 15 in the Apennines; and Apollo 17 near Littrow Crater, just beyond the southeastern edge of Mare Serenitatis.

Outstanding craters and other features

Alpine Valley A valley that cuts through the Alps, in effect linking Mare Imbrium with Mare Frigoris. It looks very much like a river valley on Earth (see page 121), but is actually a particularly straight geological fault.

Aristillus (35 miles, 56 kilometers) is one of a pair (with Autolycus) of small but prominent craters near the eastern edge of Mare Imbrium. It has a deep floor and a central range.

Aristoteles (60 miles, 97 kilometers) is one of another pair (with Eudoxus) of craters north of Mare Serenitatis.

Atlas (55 miles, 89 kilometers) and Hercules form yet another pair of craters in the north of the quadrant.

Autolycus (22 miles, 36 kilometers) forms a conspicuous pair with Aristillus.

Bessel (12 miles, 19 kilometers) is the only notable crater on Mare Serenitatis and therefore easy to spot. It is associated with a crater ray.

Cleomedes (78 miles, 126 kilometers) is a large crater just north of Mare Crisium and looks splendid just after Full Moon, along with the smaller Geminus and Messala farther north.

Eudoxus (40 miles, 64 kilometers) forms a conspicuous pair with Aristoteles.

Gauss (85 miles, 136 kilometers) is one of the largest craters in this quadrant, but is difficult to see because it is right on the limb in the northeast.

Hercules (45 miles, 72 kilometers) forms a pair with Atlas.

Hyginus Rille One of the most prominent rilles on the Moon, running from the south of Mare Vaporum toward Mare Tranquillitatis.

Manilius (22 miles, 36 kilometers), a relatively small crater on the edge of Mare Vaporum, has highly reflective walls, making it one of this quadrant's brightest spots.

Menelaus (20 miles, 32 kilometers) lies on the other side of the Haemus Mountains from Manilius and is also exceptionally bright.

Posidonius (60 miles, 96 kilometers), on the edge of Mare Serenitatis, is a beautiful crater with fascinating walls and well placed for detailed study.

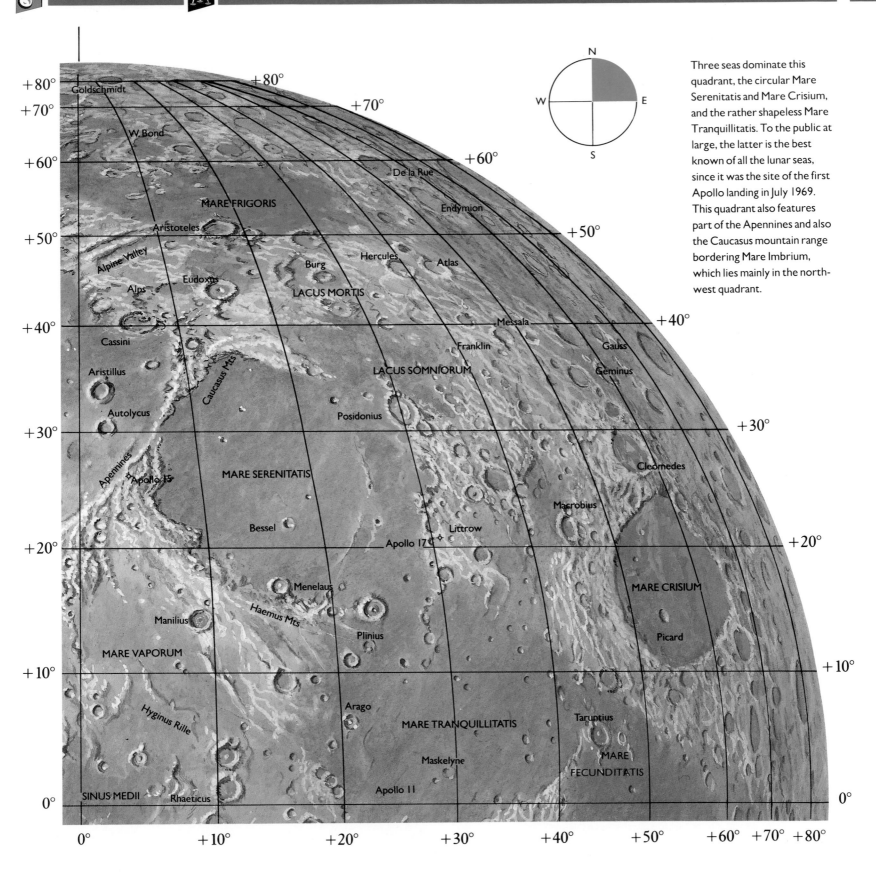

Three seas dominate this quadrant, the circular Mare Serenitatis and Mare Crisium, and the rather shapeless Mare Tranquillitatis. To the public at large, the latter is the best known of all the lunar seas, since it was the site of the first Apollo landing in July 1969. This quadrant also features part of the Apennines and also the Caucasus mountain range bordering Mare Imbrium, which lies mainly in the north-west quadrant.

Goldschmidt

W. Bond

MARE FRIGORIS

De la Rue

Endymion

Aristoteles

Alpine Valley

Burg

Hercules

Atlas

Eudoxus

Alps

LACUS MORTIS

Cassini

Messala

Franklin

Gauss

Aristillus

LACUS SOMNIORUM

Geminus

Caucasus Mts

Autolycus

Posidonius

Cleomedes

Apennines

MARE SERENITATIS

Apollo 15

Macrobius

Bessel

Littrow

Apollo 17

MARE CRISIUM

Menelaus

Manilius

Haemus Mts

Picard

Plinius

MARE VAPORUM

Hyginus Rille

Arago

Taruntius

MARE TRANQUILLITATIS

Maskelyne

MARE FECUNDITATIS

SINUS MEDII

Rhaeticus

Apollo 11

0° +10° +20° +30° +40° +50° +60° +70° +80°

SOUTHEAST QUADRANT

The irregularly shaped Mare Fecunditatis is flanked in the east by a chain of large craters, including Langrenus, Vendelinus, and Petavius.

Beyond this sea, right on the limb and straddling the equator, is Mare Smythii, which can be seen best at or shortly after Full Moon. Mare Australe, which is on the limb in the far south, can also be glimpsed at this time.

Mare Nectaris is well defined and is surrounded by a ring of large craters, including Theophilus, Catharina, and Fracastorius. Beyond this arc of craters lies the only prominent mountain range in this quadrant, the Altai Mountains, or Scarp. They run for about 300 miles (500 kilometers) from near Catharina to Piccolomini and reach a height of about 13,000 feet (4,000 meters). The range was probably formed by the same impact that created Mare Nectaris.

Outstanding craters and other features

Albategnius (80 miles, 129 kilometers) is an ancient walled plain south of Hipparchus. It shows up well at First Quarter, when it is on the terminator. It has a central range, and in the south-west has a deep crater, Klein, embedded in its walls.

Aliacensis (52 miles, 84 kilometers) is the largest of a chain of craters in the west. It forms a distinct pair with its neighbor, the slightly smaller Werner.

Fabricus (55 miles, 89 kilometers) spoils the large ruined plain Janssen and adjoins the similar-sized crater Metius.

Fracastorius (60 miles, 97 kilometers) is a badly ruined crater almost obliterated during the formation of Mare Nectaris, of which it now forms a bay.

Hipparchus (90 miles, 145 kilometers) is an ancient walled plain close to the Moon's center. It forms a pair with Albategnius, but it is more ruined. It has low walls and is not easy to spot except when on the terminator.

Langrenus (85 miles, 137 kilometers) is the most prominent of the chain of large craters that more or less follow the 60° longitude line in the east.

Maurolycus (68 miles, 109 kilometers) is an example of an ancient crater, badly eroded by later impacts.

Petavius (106 miles, 170 kilometers) One of the "big three" craters (with Langrenus and Vendelinus) on the 60° longitude line in the east. It is magnificent around Full Moon, sending out bright rays that meet those coming from Tycho in the adjacent quadrant.

Piccolomini (50 miles, 80 kilometers) lies at the southern end of the Altai range. It stands out because there are no large craters nearby.

Rheita (42 miles, 68 kilometers) is close to one of the finest crater chains on the Moon, the 100-mile (160-kilometer) long Rheita Valley.

Theophilus (62 miles, 100 kilometers) is the largest of the arc of craters west of Mare Nectaris. It has well-terraced walls and a deep floor with a central range.

Vendelinus (103 miles, 165 kilometers) lies between Langrenus and Petavius in the string of large craters in the east. It is undoubtedly more ancient than the others and has low walls. It is much less prominent than the other two.

Walter (80 miles, 129 kilometers) lies right on the 0° longitude line and is the southernmost of a string of large craters that lies just in the southeast quadrant (Ptolemaeus, Alphonsus, Arzachel, Purbach, Regiomontanus).

▼ This was part of the haul of Moon rocks brought back to Earth by the Apollo 16 astronauts. It is a typical volcanic rock containing many vesicles, or cavities, made by gas escaping when the molten rock cooled and set.

▶ Langrenus is perhaps the finest crater in this quadrant. It has high terraced walls, a central mountain range, and is spectacularly bright almost all the time.

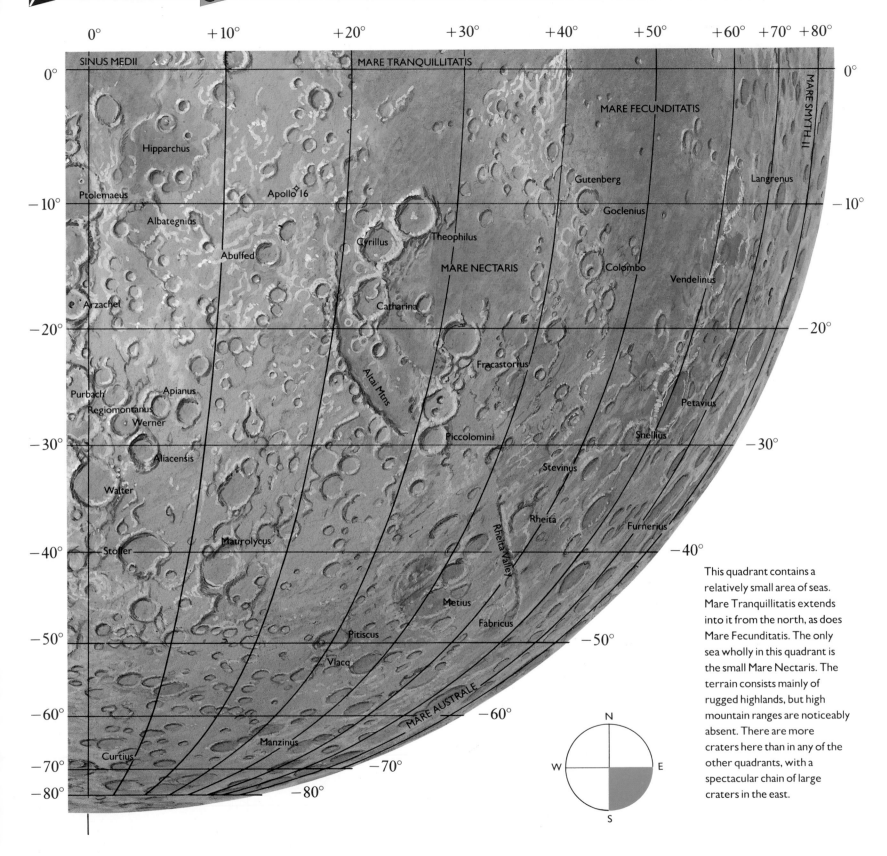

SINUS MEDII

MARE TRANQUILLITATIS

MARE FECUNDITATIS

MARE SMYTH-II

Hipparchus

Ptolemaeus

Albategnius

Apollo 16

Gutenberg

Langrenus

Goclenius

Cyrillus

Theophilus

Colombo

Abulfed

MARE NECTARIS

Arzachel

Catharina

Vendelinus

Fracastorius

Altai Mtns

Purbach

Apianus

Petavius

Regiomontanus

Werner

Piccolomini

Snellius

Aliacensis

Stevinus

Walter

Rheita

Furnerius

Stöfler

Maurolycus

Rheita Valley

Metius

Fabricus

Pitiscus

Vlacq

MARE AUSTRALE

Manzinus

Curtius

N
W   E
S

This quadrant contains a relatively small area of seas. Mare Tranquillitatis extends into it from the north, as does Mare Fecunditatis. The only sea wholly in this quadrant is the small Mare Nectaris. The terrain consists mainly of rugged highlands, but high mountain ranges are noticeably absent. There are more craters here than in any of the other quadrants, with a spectacular chain of large craters in the east.

| 1 | 2 | 3 | 4 | 5 | 6 | 7 | 8 |

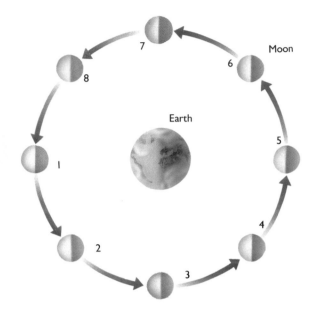

Phases of the Moon

As the Moon circles around the Earth every month, we see it go through its phases (*above*). The phases come about because, viewed from the Earth, sunlight lights up more or less of the Moon's face as it travels in its orbit.

The diagram (*left*) helps explain this. It shows the kind of view we would see if we could look down on the Earth-Moon system from above the North Pole. The numbers relate the positions of the Moon in its orbit at various times to the equivalent phase we view from Earth.

The diagrams above show how the phases come about. At New Moon phase, the Sun, the Moon, and the Earth are more or less lined up in space. The Sun lies on the opposite side of the Moon and lights up the farside. But the nearside is not lit up, so we cannot see it. After a few days, the Moon has moved far enough out of alignment for the Sun to light up a little of the near side; then we see the Moon as a crescent.

The process continues, with more and more of the near side being lit up, until the Moon is again lined up with the Sun and the Earth, but opposite the Sun in the sky. The whole near side is now lit up, and we have a Full Moon. The reverse process then takes place, with less and less of the near side illuminated, until at the next New Moon we again cannot see the near side.

Occasionally – as the Moon orbits the Earth and the Earth orbits the Sun – Sun, Moon, and Earth come into exact or almost exact alignment. Then eclipses occur. An eclipse of the Sun occurs when, at New Moon phase, the Moon's disk covers the Sun's and blots out its light (see page 128). An eclipse of the Moon occurs when at Full Moon phase, the Moon moves into the Earth's shadow.

MOON MOVEMENTS

▶ A waxing gibbous Moon, 10 days old. Note the interesting detail in the Mare Imbrium region: Sinus Iridium is nicely outlined, and the flat floor of Plato is distinctly dark. Nearby, Alpine Valley can be seen cutting through the mountains. Copernicus and its ray system are also clearly seen here.

▶ (*Opposite*) A magnificent photograph of the Moon in eclipse. The surface is lit up with a reddish hue by sunlight that has been refracted by the Earth's atmosphere, which acts like a lens. Note that we can still clearly see Tycho and some of its rays.

As regularly as clockwork, every 29½ days, the Moon goes through its phases. This time period is one of the great natural divisions of time, on which we base our calendar month (= one "moonth").

Like most other heavenly bodies, the Moon has two movements in space: it spins on its axis, and it travels in orbit around the Earth. The Moon spins around once on its axis every 27⅓ days. It also takes exactly the same period of time to travel once in orbit around the Earth. The consequence of this is that the Moon always presents the same face – the near side – to us. We say it has a "captured rotation."

The Moon's movement in orbit gives rise to its changing appearance in the sky, which we describe as its phases. It takes about two days longer for the Moon to go through its phases than to orbit the Earth. This happens because the Earth itself is moving in its orbit around the Sun.

▲ A classic lunar crater, IAU 308, located on the far side of the Moon. It has terraced walls and a central mountain range and measures about 50 miles (80 kilometers) across.

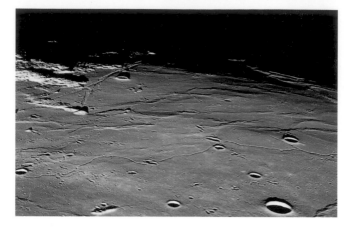

▲ The smooth surface of Mare Tranquillitatis is typical of the extensive maria on the lunar near side. Craters are few and generally small.

▼ By contrast, the surface of the lunar far side is rugged and heavily cratered. At bottom left is a small sea area, Mare Moscoviense.

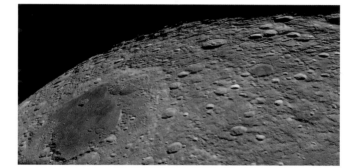

▼ This intriguing crater chain runs in a southwesterly direction from a small crater on the edge of Ptolemaeus toward the crater Davy.

THE LUNAR SURFACE

The surface of the Moon is fascinating at whatever distance we view it. The two major types of lunar landscape are the maria and the terrae – the seas and the highlands.

Both types of landscape bear witness to the bombardment the Moon has suffered from outer space in the craters that scar their surface. The floors of the maria and many of the large craters are marked by many other geological features, including rilles, ridges, faults, and domes.

The maria
On the near side of the Moon, the landscape is dominated by maria; in contrast, the far side has no maria of appreciable size. The maria have many fewer craters than the terrae, indicating that they are more recent in origin. The terrae are thought to be part of the Moon's original crust.

Analyses of rock samples brought back by the Apollo

◄ Alpine Valley, one of the Moon's most recognizable features. It cuts through the lunar Alps on the edge of Mare Imbrium and runs for about 80 miles (130 kilometers). The sinuous fault in the valley floor looks rather like a river bed.

▲ Breccia, one of the two main kinds of Moon rock, made up of pieces of old rocks cemented together.

▲ Basalt, the other main kind of Moon rock. This sample is riddled with cavities where gas escaped when the molten rock cooled.

▲ The intriguing orange soil, discovered in Taurus-Littrow Valley by the Apollo 17 astronauts.

astronauts indicate that the maria may be as young as 3.2 billion years old, while the highlands may be up to 1 billion years older.

The maria formed when massive meteorites or asteroids crashed into the Moon and gouged out huge basins. The basins were subsequently filled by vast out-pourings of molten lava. On average, the maria are several miles lower than the mean sphere of the Moon.

The craters

The craters that pepper the lunar surface were in the main also created by the impact of meteorites. The classic lunar crater has walls that rise above the surrounding surface and a floor that is lower.

One of the deepest lunar craters is the 70-mile (113-kilometer) diameter Newton, close to the south pole. The walls rise over 1.2 miles (2 kilometers) above the surrounding landscape and plunge nearly 5.6 miles (9 kilometers) below it.

When craters form, huge amounts of material are ejected from them. This "ejecta" is clearly visible around many of the younger lunar craters, such as Copernicus and Tycho. It shines brightly as crater rays, which are spectacular at Full Moon.

Material ejected from a large crater can also create lines of small craters, known as crater chains. But other kinds of crater chains may be the result of volcanic activity along a fault line in the crust.

Some of the larger craters, such as Plato and Grimaldi, are notable for their low walls and smooth, flat floor. They are rather like miniature maria and are examples of what are termed walled plains.

Rilles and ridges

Volcanic activity in the crust has also created a number of other linear features. The most prominent are the rilles, also called rimae. They are trench-like features, which can be up to about 3 miles (5 kilometers) wide and a few hundred yards deep. The winding, or sinuous, rilles are thought to be ancient lava channels, or mark where sub-surface lava tubes have collapsed. Straight rilles probably resulted when blocks of rock slipped down along fault lines.

One of the most conspicuous rilles is Hyginus Rille between Mare Vaporum and Mare Tranquillitatis. Hadley Rille, in the Apennines, was visited by Apollo 15 astronauts.

Just as there are sinuous depressions – rilles – so there are sinuous elevations. Called "wrinkle ridges," they snake across most of the maria. Some seem to be associated with volcanic activity; others seem to have formed when the maria surface cooled.

THE APOLLO EXPLORATIONS

▶ The Apollo 11 lunar module ascends into orbit to rendezvous with the CSM after the first successful lunar landing. The mare below is Mare Smythii. In the Moon sky, Planet Earth is at quarter phase.

▼ The Apollo 11 CSM in orbit, pictured from the lunar module. The landscape below is located in the north-central region of Mare Fecunditatis. The perfect symmetry of the craters at top left shows that they are relatively new.

▶ The most famous picture of the Space Age: second man-on-the-Moon Edwin Aldrin posing for a picture during the first Moon-landing mission. First man-on-the-Moon Neil Armstrong, the photographer, is reflected in Aldrin's visor. So is the Apollo 11 lunar module.

It was late evening on July 20, 1969, at the Kennedy Space Center in Florida, and early morning on July 21 in Europe, when one of the most momentous events in history took place 240,000 miles (385,000 kilometers) away.

Neil Armstrong, a being from planet Earth, stepped down onto the Moon's surface. "That's one small step for a man, one giant leap for mankind," he said.

Armstrong was the first of a dozen US astronauts who, over the next three-and-a-half years, took part in the most exciting exploration project humans have ever attempted. The project was called Apollo. It had begun in earnest following President John F. Kennedy's plea to the American people in 1961 to land a man on the Moon "before the decade is out." NASA, the National Aeronautics and Space Administration, achieved this goal for them with six months of the decade left.

The Apollo hardware

The technique chosen to land astronauts on the Moon called for a crew of three, a three-part spacecraft weighing some 50 tons (45 tonnes), and a gigantic rocket, the 365-foot (111-meter) tall Saturn V, to launch it.

The technique chosen to effect a Moon landing – lunar orbit rendezvous – involved separation and rendezvous maneuvers in lunar orbit. It required two of the crew to descend to the surface in a landing vehicle, the lunar

module, while the third crew member continued to circle overhead in the mother ship, the CSM (command and service modules). After the surface exploration, the moonwalkers returned in the top part of the lunar module and rendezvoused with the CSM, in which they all returned to Earth.

The first historic Moon landing, by Apollo 11, took

place on Mare Tranquillitatis. Armstrong and fellow moonwalker Edwin Aldrin trod the lunar soil for two-and-a-half hours, collected over 46 pounds (21 kilograms) of rock samples, and a collection of stunning photographs.

Apollo 12 (November 1969) landed on the edge of Oceanus Procellarum. During the mission, the first of five automatic scientific stations known as ALSEP (Apollo lunar surface experiments package) was set up. The Apollo 13 mission (April 1970) had to be aborted when an explosion disabled the spacecraft en route to the Moon. Apollo 14 (February 1971) set down in the region of Fra Mauro.

The remaining three missions took place in much more dramatic landscapes: Apollo 15 (July 1971) in the foothills of the Apennines, Apollo 16 (April 1972) in the highlands near Descartes Crater, and Apollo 17 (December 1972) in the Taurus-Littrow Valley near the edge of Mare Serenitatis. On all three missions, the moonwalkers were much more mobile than their earlier colleagues, for they had transportation, in the form of the lunar rover.

In all, the Apollo astronauts spent nearly 170 hours exploring the lunar surface. They traveled almost 62 miles (100 kilometers) in the lunar rover. They collected 850 pounds (385 kilograms) of rock and soil samples, and 50 core samples. They took 30,000 photographs on the surface and from lunar orbit, and acquired all kinds of data on 20,000 reels of magnetic tape.

Scientifically and technologically, Apollo was an outstanding triumph. Posterity will have to judge whether it was worth the 25 billion dollars it cost.

▲ Geologist Harrison Schmitt works near a huge split boulder at Taurus-Littrow on the Apollo 17 mission (the last).

▶ Mount Hadley provides the backdrop as James Irwin salutes the "Stars and Stripes" at the start of the Apollo 15 mission to the foothills of the Apennines. At right is the lunar rover, being used on the Moon for the first time.

◀ Edwin Aldrin struggling with a core-sampling drill on the flat landscape of Mare Tranquillitatis during the historic first Moon landing. The core samples give clues to the history of the lunar surface.

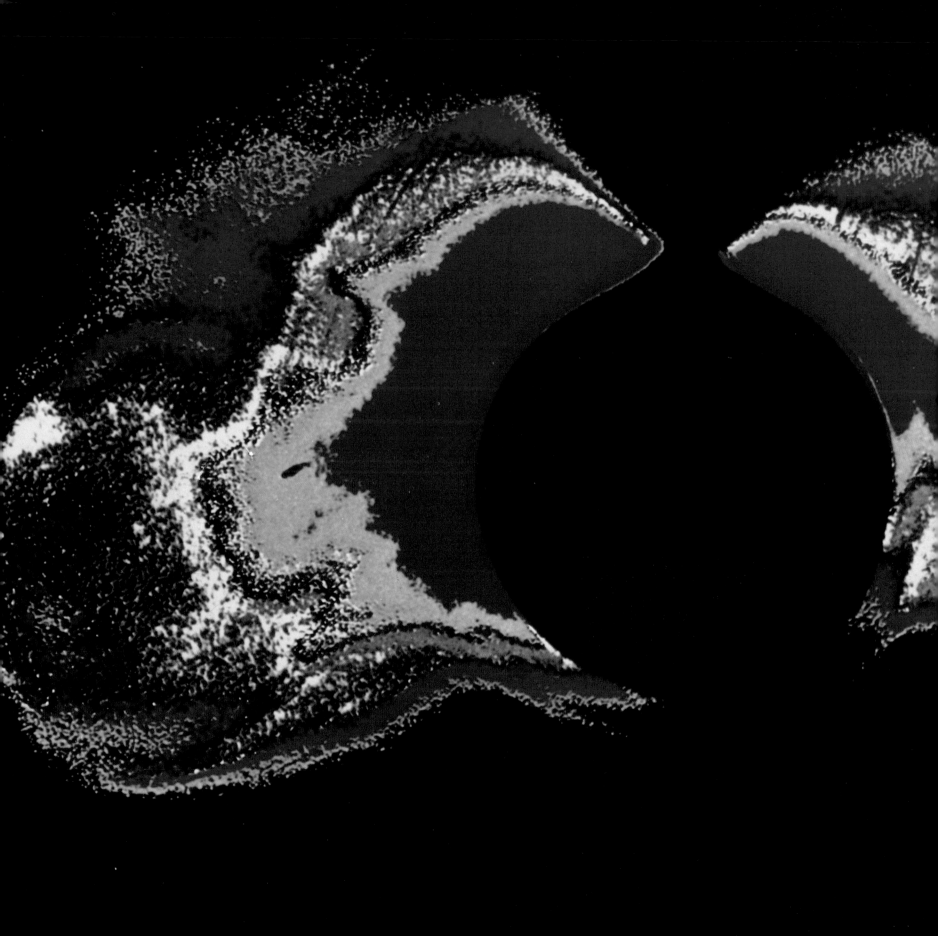

◄ This false-color picture shows the Sun's corona, its outer atmosphere of rarified gases that extends for millions of miles in all directions before it merges imperceptibly into interplanetary space. This was one of the incomparable pictures taken in 1973 during the Skylab project.

5

THE SUN

Asked the question "Which is the nearest star?," we would probably reply Proxima Centauri, a faint red dwarf more than 4 light-years away. But, strictly speaking, this is wrong. The nearest star to us is a quite bright yellow dwarf, only 8.5 light-minutes away – the daytime star we call the Sun.

This golden orb, which arcs through our skies by day, pours out the light and heat that makes life on Earth possible in its myriad of different forms. Astronomers study the Sun closely, since its behavior mirrors that of the majority of the stars in the universe.

But the amateur astronomer must study the Sun with care, since its blinding light will indeed blind. It must never be observed directly through binoculars or telescopes. However, these instruments can be used to project solar images, which will reveal such phenomena as sunspots and eclipses.

Total eclipses of the Sun are phenomena that many astronomers rave over and journey to the ends of the Earth to see. And for good reason. They are among the most beautiful, the most exciting, and the most awesome happenings in nature.

▲ The golden orb of the Sun, about to sink below the horizon at Cape Canaveral in Florida. At sunset the sky turns reddish-orange because dust in the lower layers of the atmosphere blocks the shorter wavelengths in sunlight.

▼ Impromptu solar astronomy in the Australian outback in March 1986, when Halley's Comet chasers pause to witness a partial solar eclipse. A double image of the partially eclipsed Sun is obtained by projection through binoculars.

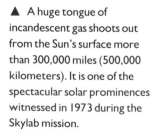

▲ A huge tongue of incandescent gas shoots out from the Sun's surface more than 300,000 miles (500,000 kilometers). It is one of the spectacular solar prominences witnessed in 1973 during the Skylab mission.

DAYTIME STAR

Our daytime star, the Sun, is a huge globe of hot gas, about 865,000 miles (1,390,000 kilometers) in diameter. It is a very average star, much bigger than some, but much smaller than others.

The Sun is 750 times more massive than all the other bodies in the solar system put together. With such a huge mass, it has very powerful gravity, which keeps these bodies captive.

The Sun lies at an average distance from Earth of about 93 million miles (150 million kilometers). We call this distance one astronomical unit (AU).

The Sun pours out into space a fantastic amount of energy, not only as light and heat (infrared rays), but also as many other kinds of radiation, such as gamma rays, X-rays, ultraviolet rays, and radio waves. It has poured out this energy since it was born, some 4,600 million years ago. And it should continue to do so for another 5,000 million years. This energy is produced by nuclear fusion reactions taking place in the hydrogen gas in the Sun's core.

Solar energy radiates away from the visible surface of the Sun, which we call the photosphere ("light-sphere").

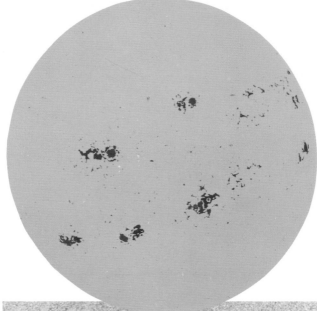

or so. And they come and go according to a "sunspot cycle" of about 11 years. We can study sunspots using a telescope to project an image of the Sun onto a white card. **Never look directly at the Sun through a telescope or binoculars**.

Often associated with sunspots are eruptions called flares. They give rise to streams of charged particles, which greatly increase the solar wind, the normal outflow of particles from the Sun.

It is at such times, when the solar wind "blows" exceptionally strongly, that magnetic storms and aurorae occur on Earth. Aurorae are visible mainly in far northern and far southern skies, where they are known respectively as the aurora borealis ("Northern Lights") and aurora australis ("Southern Lights"). They take the form of lovely glows and shimmering curtains of colored light. The light is given off when the charged particles of the solar wind collide with air particles high in the atmosphere.

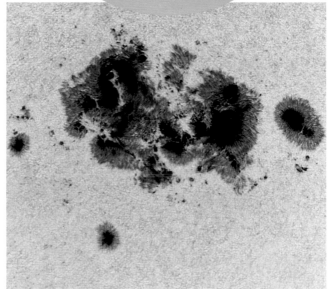

◄ A well-developed sunspot group. Note in each spot the dark umbra at the center and the surrounding lighter penumbra. Some sunspots can grow to more than 125,000 miles (200,000 kilometers) across and persist for months.

◄ (Top) Sunspots are the visible signs of violent activity taking place on the Sun's surface.

▼ The aurora borealis, viewed from Kitt Peak National Observatory, Arizona, in April 1981. It is unusual for vivid aurorae to occur at such low latitudes and indicates that violent activity was taking place on the Sun at the time.

Its temperature is about 11,000°F (6,000°C). Above the photosphere is the Sun's lower atmosphere, called the chromosphere ("color-sphere") because of its reddish color. And above that is the outer atmosphere, the corona, which becomes increasingly more rarified until it merges into the near-vacuum of space. Because the photosphere is so bright, we can see the chromosphere and corona only during a total eclipse of the Sun.

The Sun's surface is in constant turmoil as hot gases well up from below. Sometimes great fountains of flaming gas called prominences shoot high above the surface. From time to time, dark regions develop and grow. We call these regions sunspots. They appear to move across the solar disk as the Sun rotates on its axis once every 25 days

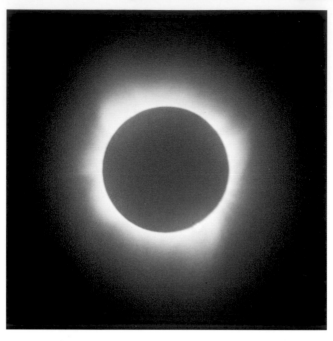

▼ A total solar eclipse occurs when the Moon's shadow falls on the Earth.

OBSERVING SOLAR ECLIPSES

▲ The author's record of the March 18, 1988, total eclipse, viewed from Bangka Island, Indonesia. (*Left*) Seconds before totality, the Moon's shadow is clearly visible. (*Right*) During totality, the corona is brilliant.

▼ The author, appropriately attired for viewing the long total eclipse from Hawaii (July 11, 1991)! Note the simplicity of the equipment: ordinary camera with telephoto lens and mylar filter, cable shutter release, tripod.

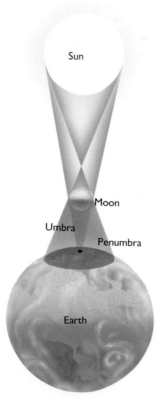

Of all the natural phenomena we witness on Earth, none is more spectacular or more awe-inspiring than a total eclipse of the Sun. During an eclipse, day turns suddenly into night; the air chills; birds stop singing and start to roost.

Eclipses of the Sun occur because of a strange astronomical coincidence. The Sun is four hundred times farther away from the Earth than the Moon, but its diameter is four hundred times greater than that of the Moon. The consequence of this is that both Sun and Moon appear much the same size in the Earth sky.

From time to time, as the Moon travels in orbit around the Earth, it lines up exactly with the Sun and the Earth. When it comes between them, it blots out the Sun's light and casts a shadow on the Earth. Then we have an eclipse of the Sun, or a solar eclipse. When the Moon is on the other side of the Earth from the Sun, it enters the shadow cast by the Earth, and we have an eclipse of the Moon, or a lunar eclipse (see pages 118 and 119).

On average, solar eclipses occur about twice a year, but they are not always total; they are often partial, the Moon covering only part of the Sun's disk. Sometimes they are annular, with the Moon covering up all of the

Sun except for an annulus, or ring, around the edge.

Even under the most favorable conditions during a total eclipse, the diameter of the umbra, or complete shadow, cast on the Earth by the Moon is seldom more than about 170 miles (270 kilometers). This shadow sweeps out a "path of totality" as it races across the Earth at speeds of several thousand miles an hour. The maximum duration of totality occurs in the middle of the path, along the center line. The longest any total eclipse can occur is theoretically about seven-and-a-half minutes. But over the next 50 years, it is predicted that the longest (in July 2009) will be 6 minutes 38 seconds. Most will be much shorter.

Here is a brief summary of total eclipse "events." The Moon takes on average about two hours to pass across the Sun. The partial phase of the eclipse begins at "first contact," as the Moon takes its first "bite" out of the Sun's disk. About an hour later, with the light fading perceptibly all the while, comes "second contact," the beginning of totality. Just before this comes the phenomenon of "Baily's Beads," as beads of sunlight shine through the valleys on the limb of the Moon.

During totality, vivid pink prominences show up around the dark disk of the Moon. Then the Sun's pearly white corona appears. More Baily's Beads precede the end of totality ("third contact"). Then the Sun emerges once more, creating for an instant a beautiful "diamond ring." About an hour later, the Moon completely clears the Sun ("fourth contact"), and the eclipse is over.

For the total eclipse of July 11, 1991, eclipse-watchers – the author among them – descended on the Big Island of Hawaii in their thousands. At about 7.28 in the morning, the Sun went into total eclipse for the relatively long period of 4 minutes 12 seconds. The author witnessed the event from Kailua Kona, close to the center line of the path of totality. Clouds threatened to mar the occasion, but fortuitously rolled away minutes before totality. Thousands of others on the island were not so lucky.

▼ During totality, the sky is not completely dark. It is tinged orange on the horizon.

► The Moon has already taken a significant "bite" out of the Sun. Cloud is creating a halo.

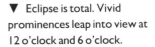

▼ Eclipse is total. Vivid prominences leap into view at 12 o'clock and 6 o'clock.

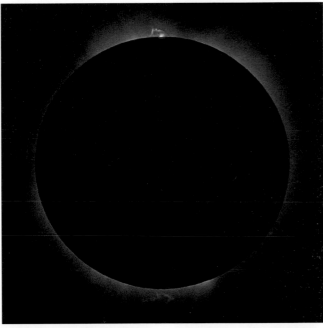

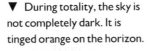

▲ An exquisite "diamond ring" signals the end of totality; daylight is about to return.

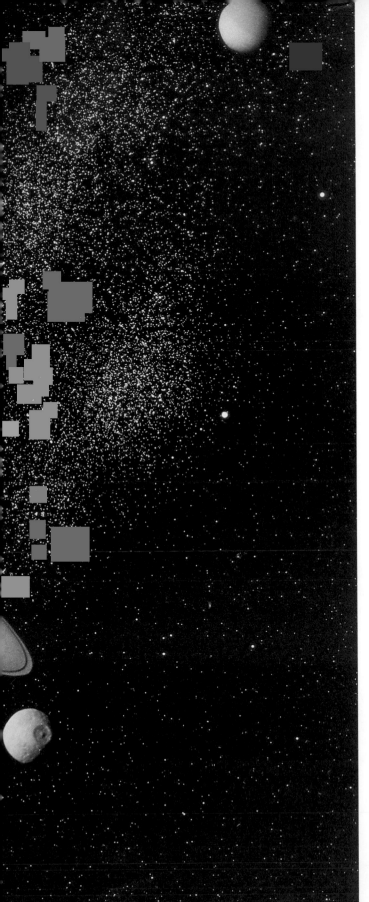

◀ The ringed planet Saturn, pictured here with six of its moons against a stellar background. It is a montage of photographs taken by the Voyager I space probe. The moon in the foreground is Dione; then, going clockwise, come Enceladus, Rhea, Titan, Mimas, and Tethys.

6

THE SUN'S FAMILY

The Sun is really massive and consequently has an enormous gravitational pull, which reaches out thousands of millions of miles into space. It holds together a varied collection of heavenly bodies, large and small. They form the Sun's family, or Solar System.

The Earth is one of the larger bodies in the solar system that we call planets. There are eight more planets, four smaller than the Earth and four very much larger. Three of the other planets in particular – Venus, Mars, and Jupiter – far outshine the brightest stars and are easy to follow as they wander across the celestial sphere in and out of the constellations month by month.

In addition to planets, the Sun's family includes more than 60 moons, thousands of mini-planets, or asteroids, and unknown numbers of icy lumps, which become visible as comets when they approach the Sun. Even specks of interplanetary dust reveal their presence in the night sky when they burn up in the atmosphere, creating meteors or shooting stars. All these bodies make rewarding study, with the naked eye as well as with binoculars and telescopes.

THE SOLAR SYSTEM

Until the sixteenth century, people believed that the Earth was fixed at the center of the Universe, and that all the heavenly bodies circled around it. This followed the system laid down by the Greek astronomer Ptolemy in about AD 150.

▼ The planets are the major bodies in the solar system. They circle around it at different distances, but in much the same plane. Viewed from the "north," the planets travel counterclockwise around the Sun. All except Venus also spin counterclockwise on their own axis. In this diagram the orbits of the planets are drawn roughly to scale.

Some astronomers were dissatisfied with this belief, which could not readily account for the observed motions of the planets. One was Copernicus, who realized that planetary motions could be explained if the Sun and not the Earth were at the center of the Universe. Because he relegated the Earth to the status of a mere planet, the Church regarded Copernicus's views as heresy. They were not generally accepted until early the next century, when Kepler worked out his celebrated laws of planetary motion.

Copernicus had no idea of the scale of the solar system, nor had any astronomers until comparatively recent times. We know now that it reaches out from the Sun far beyond the Earth and indeed far beyond the orbit of Pluto, usually the most distant planet we know.

Pluto wanders at times over 4.5 billion miles (7 billion kilometers) from the Sun. But way beyond the path of that maverick planet lurks a "belt" of icy lumps (the Kuiper Belt) and beyond that a similarly composed "cloud" (the Oort Cloud). This "cloud" is thought to be a reservoir of icy bodies that occasionally stray into the inner solar system and become visible as comets. It is thought that the Oort Cloud could reach out more than two light years from the Sun – halfway to the nearest stars. The existence of such far-distant bodies seems to have been confirmed by the discovery in 1992 of an object (QB1) beyond Pluto's orbit. Some 125 miles (200 kilometers) across, it could be part of the Kuiper Belt.

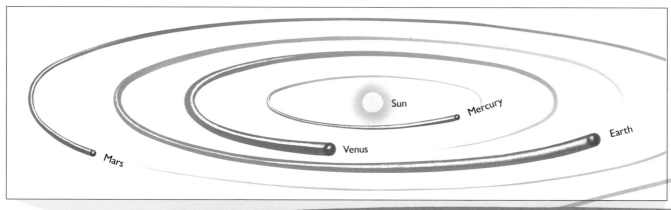

◄ This inset diagram shows the orbits of the four inner planets to scale. Mercury, Venus, and Mars are called the terrestrial planets because they are rocky bodies like the Earth.

Planet		Mercury	Venus	Earth	Mars	Jupiter	Saturn	Uranus	Neptune	Pluto
Diameter at equator	(km)	4,878	12,104	12,756	6,794	142,800	120,000	51,200	49,500	2,280
	(miles)	3,031	7,521	7,926	4,222	88,700	74,600	31,800	30,800	1,430
Mean distance from Sun ($\times 10^6$)	(km)	60	108	150	228	778	1,430	2,870	4,500	5,900
	(miles)	37.5	67.5	93.5	143	486	890	1,790	2,810	3,690
Circles Sun in		88 days	225 days	365.25 days	687 days	11.9 years	29.5 years	84 years	165 years	248 years
Spins on axis in		58.6 days	243 days	23.93 hours	24.6 hours	9.8 hours	10.2 hours	16.3 hours	16 hours	6.3 days
Mass (Earth = 1)		0.06	0.82	1.00	0.11	318	95	14.5	17.2	0.002
Volume (Earth = 1)		0.05	0.88	1.00	0.15	1,316	755	52	44	0.005
Density (water = 1)		5.4	5.2	5.5	3.9	1.3	0.7	1.3	1.8	2
Number of moons		0	0	1	2	16+	22+	15+	8+	1

▼ Pluto's orbit takes it a long way above and below the other planets. The orbit is highly eccentric and at times takes Pluto inside the orbit of Neptune. It is inside Neptune's orbit at present and will remain there until 1999. Then it will resume its rightful place as the most distant planet.

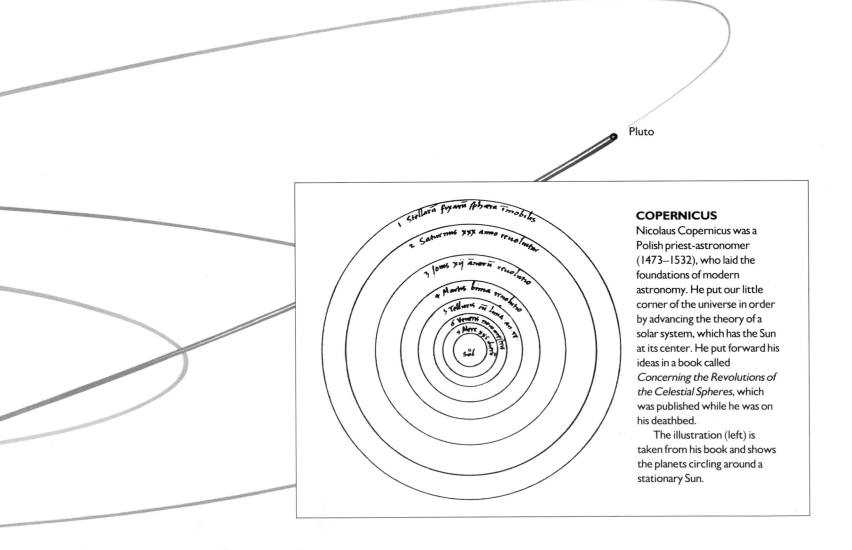

Pluto

COPERNICUS

Nicolaus Copernicus was a Polish priest-astronomer (1473–1532), who laid the foundations of modern astronomy. He put our little corner of the universe in order by advancing the theory of a solar system, which has the Sun at its center. He put forward his ideas in a book called *Concerning the Revolutions of the Celestial Spheres*, which was published while he was on his deathbed.

The illustration (left) is taken from his book and shows the planets circling around a stationary Sun.

▲ Three views of Mars during the opposition of 1967. Note the north polar ice cap, the whitish haze in places, and various dark markings.

▼ (*Right*) Saturn through a telescope. The "A" and "B" rings stand out clearly.

(*Middle*) Jupiter through a telescope, showing prominent belts and zones and a vivid Great Red Spot.

(*Left*) Jupiter shining brightly alongside a cloud-shrouded Full Moon.

OBSERVING PLANETS AND ASTEROIDS

Five planets – Mercury, Venus, Mars, Jupiter, and Saturn – were known to the ancients, for they can be seen with the naked eye. When most favorably placed for observation, they rival or outshine even the brightest stars. They show up in telescopes as distinct disks rather than points of light.

From the amateur's point of view, these five planets most reward observation. The others – Uranus, Neptune, and Pluto – are too remote to show anything of interest.

Because of their brilliance, Venus, Jupiter, and often Mars are easy to spot in the night sky. Mercury and Saturn are rather more elusive. You can find out exactly where the planets are at any time by consulting astronomical magazines and yearbooks.

The motion of the planets through the heavens is not entirely straightforward. For example, Mars, Jupiter, and Saturn, which normally travel eastward against the background of stars, sometimes start traveling westward, or in a retrograde direction. This happens because the Earth, traveling more quickly inside their orbit, periodically catches up with and then overtakes them. And they appear to "loop the loop" in the sky.

Mercury and Venus
Although Mercury is not very far away, it is not easy to pick up. This is because it orbits relatively close to the Sun and therefore stays near it in the sky. At its most favorable, Mercury can reach magnitude –1.9. Venus is

very much easier to see. It is much nearer than Mercury and much bigger, and at maximum brightness reaches a magnitude of –4.4.

When either of these planets lies east of the Sun, they are visible in the west after sunset. When they lie west of the Sun, they are visible in the east before sunrise. In other words, they are either evening or morning stars.

Mars

This planet, when most favorably placed, reaches a magnitude of –2.8, outshining all the other planets except Venus. It can readily be distinguished from the others because of its orange-red color, which earns it the name of the Red Planet.

Whereas little detail can be seen on the disk of Venus and Mercury, Mars is more revealing. It shows prominent ice caps at the poles and dark markings that come and go with the changing seasons.

Asteroids, or minor planets

These mini-planets orbit in a broad "belt" between the orbits of Mars and Jupiter. The largest ones – the biggest is Ceres, about 620 miles (1,000 kilometers) across – can be glimpsed and photographed moving against the background of stars (*see below*).

Jupiter and Saturn

Jupiter is by far the largest planet, and it is covered with clouds that reflect sunlight well. This adds up to it rivalling Mars in brightness, reaching up to magnitude –2.6. Jupiter is easily distinguished because it is brilliant white. Larger telescopes will show quite a lot of detail on Jupiter's disk, particularly the dark and light bands ("belts" and "zones"), and maybe the famous Great Red Spot.

Saturn's disk is a pale imitation of Jupiter's, but this planet's shining glory is the incredibly beautiful ring system that girdles its equator. The aspect of the rings changes noticeably during Saturn's near 30-year orbit around the Sun.

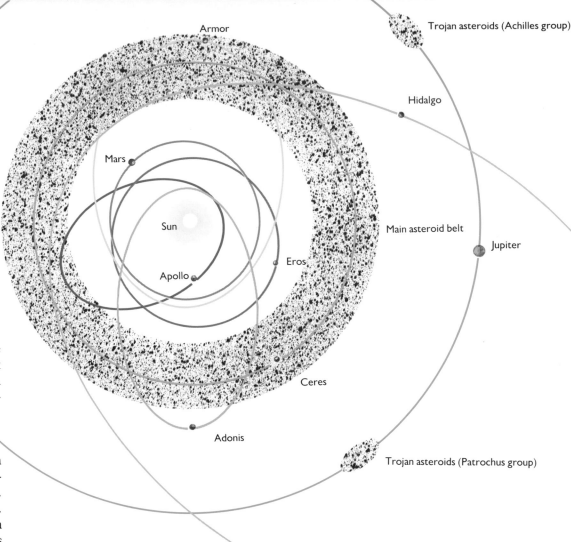

▲ The asteroid belt lies between about 170 and 340 million miles (275 and 550 million kilometers) from the Sun between the orbits of Mars and Jupiter. Some of the asteroids stray far away from the main belt, as the diagram shows. Two groups of asteroids, the Trojans, travel in the same orbit as Jupiter, locked in place by the planet's gravity.

◄ Asteroids appear on long-exposure star photographs as trails because they are moving against the stellar background. In stellar photography the telescope is driven to follow the movement of the stars.

OBSERVING METEORS AND COMETS

You don't have to gaze at the night sky for long before you see luminous streaks shooting across the sky. We call them meteors, or shooting stars. They are not stars, of course, but fiery trails left behind when particles from outer space burn up high in the atmosphere.

The space between the planets in the solar system is not completely empty, but contains significant amounts of particles of rock and metal. These particles are known as meteoroids. Most are no bigger than a grain of sand, but some can be the size of large boulders.

As the Earth travels in space, it encounters meteoroid particles all the time. When they get near the Earth, they are attracted to it by gravity. They plunge into the upper atmosphere at speeds of up to 43 miles (70 kilometers) a second. Friction with the air particles generates enough heat to make them glow white-hot and eventually burn up. They leave behind the fiery trail of glowing gases we see as a meteor.

Fireballs and meteorites

Sometimes an exceptionally big meteoroid smashes into the atmosphere. It appears in the night sky as a great ball of fire we call a fireball, or bolide. If the meteoroid causing the fireball is big enough, it may survive to reach the ground, as a meteorite. It may even gouge out a crater.

Two basic types of meteorite are found, stony and iron. The stony ones are made up mainly of silicates, like many stones on Earth. The iron ones are made up of a mixture of iron and nickel, sometimes with a little cobalt.

At certain times of the year, the number of meteors increases dramatically from one about every 10 minutes to maybe one or more a minute. This happens during so-called meteor showers. They occur when the Earth passes through trails of dust left behind by passing comets (*see below*).

All the meteors appear to come from one point in the sky, called the radiant. The showers are named after the constellation containing the radiant. One of the most reliable showers, the Perseids, takes place in August each year. The Leonids (November) and the Geminids (December) can also be impressive.

Comets

Comets appear less often in our skies, but can become the most spectacular of heavenly objects. The majority appear without warning, traveling into the inner reaches of the solar system from a "reservoir" (the Oort Cloud) far beyond the orbits of the planets. Some comets, called the periodic comets, reappear regularly in the Earth's skies, traveling to and fro in known orbits. The best-known of them all, Halley's Comet, returns about every 75 years. The comet with the shortest period is Encke's Comet, which takes only 3.3 years to orbit the Sun.

Comets are lumps of ice and dust, normally frozen solid. Typically they are a few tens of miles across. For most of the time they remain remote and invisible. Only

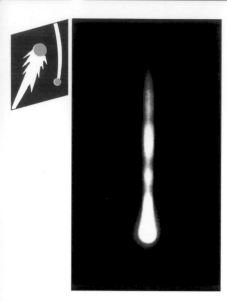

▲ A fireball leaves a fiery trail as it streaks through the night sky. Objects producing fireballs may be large enough to reach the ground as meteorites.

▼ Meteor Crater in Arizona is remarkably well preserved, as this picture testifies. It measures 4,150 feet (1,265 meters) across and is 575 feet (175 meters) deep.

◄ Halley's Comet, pictured against a starry backcloth, on April 9, 1986, on its last return to Earth's skies. It will not return again until 2061.

► Comet Ikeya-Seki (1965), visible in broad daylight in some parts of the world, was one of the brightest comets of this century.

▼ On the night of March 13/ 14, 1986, the European space probe Giotto returned this image as it homed in on Halley's Comet. The nucleus, the darker region at top left, appears to be about 9 miles (15 kilometers) long and about 5 miles (8 kilometers) across.

when they travel into the inner solar system do they make their presence known. The Sun's heat begins to evaporate some of the ice, which releases some dust. This vapor and dust cloud reflects sunlight, so the comet becomes visible. The "pressure" of the solar wind "blows" the gas and dust cloud away from the head of the comet to form a tail, which can stretch for millions of miles.

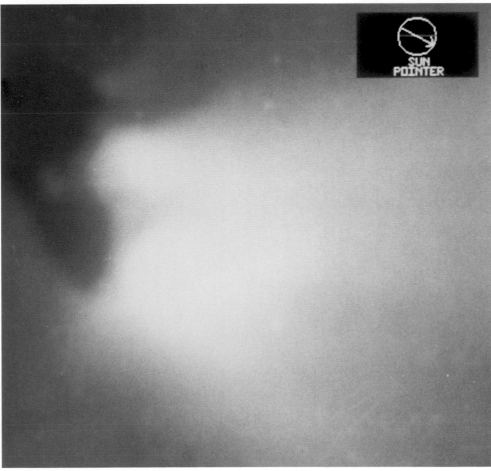

▼ Comets pursue quite different orbits from the planets, which all circle the Sun in much the same plane. Comets approach and leave the inner Solar System at all angles.

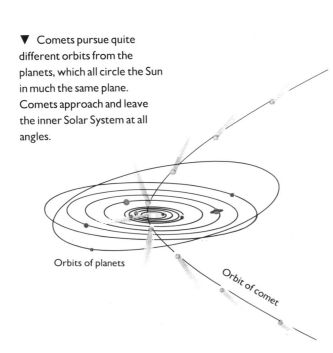

Orbits of planets

Orbit of comet

► The surface of Mercury is heavily cratered and bears more than a passing resemblance to that of the Moon. The craters are similar to Moon craters, many of them having the typical central mountain peak. Note the prominent fault line, or scarp, near the limb.

▼ This oblique relief view (with false color) of the surface of Venus was computer-generated from data returned by the Magellan space probe. The probe, launched in May 1989, mapped virtually the whole planet over a four-year period. It found hundreds of volcanoes, with evidence of extensive lava flows, and spider-like formations called arachnoids, which are unique to Venus.

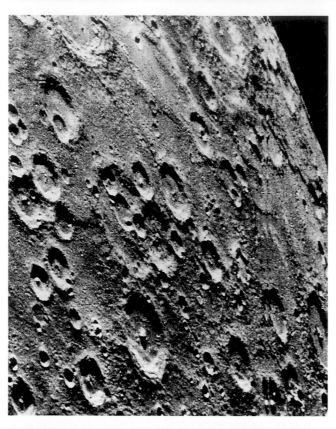

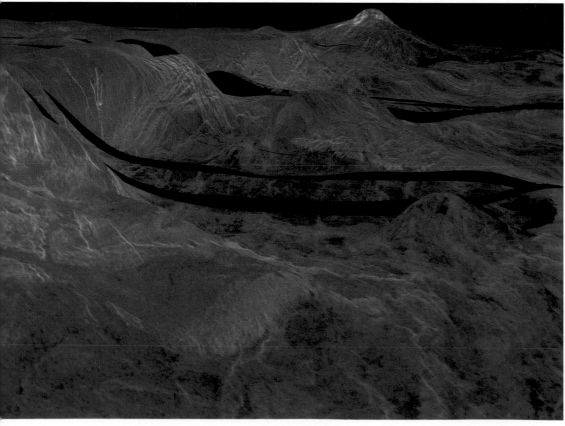

EARTHLIKE PLANETS

The three planets that lie relatively close to Earth in the inner reaches of the solar system are quite different from the ones that lie much farther out.

These planets – Mercury, Venus, and Mars – are relatively small and made up mainly of rock, like the Earth, which is why they are often called the terrestrial planets. They contrast markedly with the outer planets, which are giant balls of gas.

Mercury

Second smallest of the planets after Pluto, Mercury (diameter 3,031 miles, 4,878 kilometers) is the one closest to the Sun. It rotates slowly, taking 59 days to spin around once, while it circles the Sun every 88 days.

This slow rotation results in a point on the surface facing the Sun for periods of three months at a time. Then, surface temperatures rise as high as 800°F (425°C), which is hot enough to melt lead. On the other hand, a point on the opposite side of Mercury receives no sunshine at all for three months, and there temperatures can fall to −275°F (−170°C).

Mercury is little bigger than the Moon and likewise has no atmosphere. Its surface is also covered with craters. Unlike the Moon, however, Mercury has no large seas, or maria. Its most distinctive large surface feature is the ring-shaped Caloris Basin, which measures about 870 miles (1,400 kilometers) across.

Venus

Only marginally smaller than the Earth, Venus (diameter 7,521 miles, 12,104 kilometers) takes about 225 days to circle the Sun. It takes even longer, 243 days, to spin once on its axis. It also spins around in the opposite direction from all the other planets.

Although a near-twin of the Earth in size, Venus could hardly be more different in other respects. It has a heavy atmosphere, consisting mainly of carbon dioxide at a crushing pressure 90 times the atmospheric pressure on Earth. A runaway "greenhouse effect" has raised temperatures at the surface to 890°F (475°C) or more.

Thick clouds in the atmosphere, some formed of sulfuric acid droplets, keep the surface of Venus permanently hidden from view. We can "see" the surface only by radar. Radar scans have revealed that most of the planet consists of rolling plains. There are just two highland regions – Aphrodite Terra, which is about the size of Africa, and Ishtar Terra, which is smaller.

Mars, the Red Planet

Although it is much smaller, Mars (diameter 4,222 miles, 6,794 kilometers) resembles the Earth in a number of ways. It spins on its axis in about the same time; it has seasons; it has a slight atmosphere; and it has polar ice caps, which advance and retreat in turn in winter and summer. Because of such similarities, people once thought that Mars might harbor life, even intelligent life. But investigations by space probes such as Viking have ruled this out.

Mars takes about 687 days to circle the Sun. It is a much colder world than the Earth, with temperatures struggling to reach freezing point even at the equator in summer. In winter, temperatures can drop to −220°F (−140°C). Carbon dioxide is the main gas in the very thin atmosphere, which has only one-hundredth the pressure of Earth's atmosphere.

Although there is no running water on Mars, there is water ice at the poles and traces of moisture in the atmosphere. This leads to clouds, mist, and frost occurring in places at times.

Parts of Mars are quite heavily cratered, but its most prominent features include a huge canyon and four massive volcanoes. The Martian "Grand Canyon," Valles Marineris (Mariner Valley), runs for nearly 3,000 miles (5,000 kilometers) near the Martian equator and averages more than 60 miles (100 kilometres) across. Mariner Valley is not far from the four prominent volcanoes. The highest is the 15-mile (25-kilometer) high Olympus Mons.

▲ The crater complex at the summit of Olympus Mons, a huge extinct volcano on Mars.

◄ This Viking approach picture shows several of Mars's outstanding features. The line of circles near the center shows the high volcanoes on the Tharsis Ridge. Above them is the massive crater of Olympus Mons. At the bottom of the picture is the circular Argyre Basin, filled with frost.

► Rust-red rocks are strewn across the Martian plain of Chryse, landing site of the first Viking lander in 1976.

THE GIANT PLANETS

The terrestrial planets are midgets compared with the four giant planets that lie in the outer reaches of the solar system. In order of increasing distance from the Sun, they are Jupiter, Saturn, Uranus, and Neptune.

With a diameter 11 times that of Earth, Jupiter is the giant among giants and has a greater mass than all the other planets put together! Yet even it is a dwarf compared with the Sun.

Of the four giants, only Jupiter and Saturn were known to ancient astronomers. The other two are too faint to be seen except in telescopes. William Herschel discovered Uranus in 1781, while Johann Galle discovered Neptune in 1846.

Thanks to NASA space probes, particularly the two Voyagers, all four giants are now as familiar to us as the four terrestrial planets. The giant planets differ markedly from the Earth-like inner planets in composition. They are made up mainly of the two gases, hydrogen and

helium. They all have deep atmospheres, and deep oceans of liquid gas underneath. Only at their centers might there be a rocky core.

Jupiter

The light and dark belts and zones we see on Jupiter (diameter 88,700 miles, 142,800 kilometers) through a telescope are bands of clouds drawn out parallel by the planet's swift rotation – it spins around once in under 10 hours. Jupiter takes nearly 12 years to make one orbit of the Sun.

Jupiter's disk is brilliantly colored red and orange. The furious rotation of the atmosphere sets up turbulence, particularly in the region of the huge storm center we know as the Great Red Spot, which measures about 17,000 miles (28,000 kilometers) across.

Saturn

The second largest planet, Saturn (diameter 74,600 miles, 120,000 kilometers) is also the least dense. With a relative density of less than 1, it would, if it could, float in water. Saturn's disk is less obviously banded than Jupiter's, though winds on the planet blow more strongly – up to 1,100 mph (1,800 km/h). Saturn circles the Sun once every 29½ years.

The glorious ring system around Saturn's equator spans more than twice the planet's diameter, but is at most only about 0.6 mile (1 kilometer) thick. From Earth only three main rings are readily visible: A, B, and C. Space probes have found several more and have discovered that they are made up of thousands of individual ringlets. These mark the paths of swift-moving rock and ice particles.

Uranus and Neptune

These two planets are near-twins in size; Uranus (diameter 31,800 miles, 51,200 kilometers) is about 1,000 miles (1,600 kilometers) bigger across than Neptune. At present, Neptune is the most distant of the planets because Pluto, usually the most distant, is traveling inside Neptune's orbit (until 1999). Uranus takes just slightly over 84 years to complete one orbit of the Sun, while Neptune takes almost twice as long – nearly 165 years.

The most unusual feature of Uranus is that its axis is tilted at an angle of 98° – more than a right-angle – with respect to the plane of its orbit. In effect, the planet travels on its side as it orbits the Sun.

The disk of Uranus is a bluish-green color and shows virtually no features. The disk of Neptune is deep blue and is flecked with wisps of white clouds, particularly around the dark spots, which are storm centers. Voyager 2 spotted faint rings around both Uranus and Neptune.

▼ Colorful clouds are drawn into parallel bands by the swift rotation of Jupiter's atmosphere. Here, in the south equatorial belt, the clouds eddy furiously around the Great Red Spot.

◀ This is the view you would see if you were orbiting Uranus's moon, Miranda. The blue-green colored planet is about 81,000 miles (130,000 kilometers) distant. Below, Miranda's contorted surface is geologically the most baffling in the whole solar system.

▼ Visually, Neptune is much more interesting than Uranus. It has a surprising amount of "weather" for a planet so far from the Sun. It features stormy spots and banks of clouds. Of particular interest is the Great Dark Spot shown in the picture, a massive storm center analogous to Jupiter's Great Red Spot.

▲ A beautiful Voyager picture of the ringed planet, Saturn. It shows clearly two of the three main rings: A (outer) and B, separated by the Cassini Division. The bands of clouds in the atmosphere are less prominent than Jupiter's.

▼ Saturn's extensive ring system, compared in size with the Earth. In all, it measures some 170,000 miles (270,000 kilometers) across. The different colors in this false-color image of the rings indicate mainly different sizes of ringlet particles.

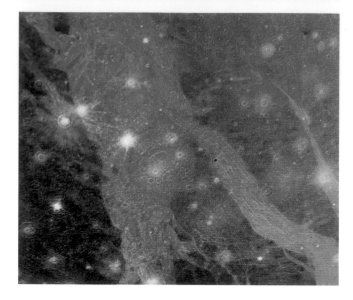

▲ A close-up of Ganymede's surface. The white spots are fresh craters, made by impacts with rocky debris. The impacts have gouged out holes and exposed fresh ice. The curious grooved terrain shows where the crust has fractured.

▼ A volcano erupting on the limb of Io. It is spewing out molten sulfur rather than lava, which accounts for the vivid yellow-orange color of the surface. Gas and dust are being blasted to heights of 125 miles (200 kilometers) or more.

▲ The largest moons in the solar system compared in size. With the exception of Titan, all the moons in the outer circle belong to Jupiter. Titan is one of Saturn's moons. Ganymede (3,278 miles, 5,275 kilometers) is the largest moon, followed in turn by Titan (3,200 miles, 5,150 kilometers), Callisto (2,995 miles, 4,820 kilometers), Io (2,257 miles, 3,632 kilometers), the Moon (2,160 miles, 3,476 kilometers) and Europa (1,942 miles, 3,126 kilometers).

MANY MOONS

All the planets except Mercury and Venus have one or more satellites, or moons, orbiting around them. These bodies range in size from irregular lumps of rock tens of miles across to bodies bigger than Mercury.

The Earth, of course, has just one moon – the Moon. Pluto also has only one moon, Charon, which, amazingly, is half its size, with a diameter of about 740 miles (1,190 kilometers). Mars has two tiny moons – Phobos (about 17 miles, 28 kilometers across) and the slightly smaller Deimos. It is almost certain that these two rocky bodies are asteroids captured from the nearby asteroid belt.

The giant planets have an astonishing number of moons between them – more than 60! They are made up mainly of ice and rock. The Voyager probes discovered many small new moons when they flew past the planets in the 1970s and 1980s.

Jupiter has at least 16 moons, the four largest being the Galilean moons, which we can see in binoculars – Io, Europa, Ganymede, and Callisto. Of these, the most interesting is Io, the only body we know, apart from the Earth, that has active volcanoes.

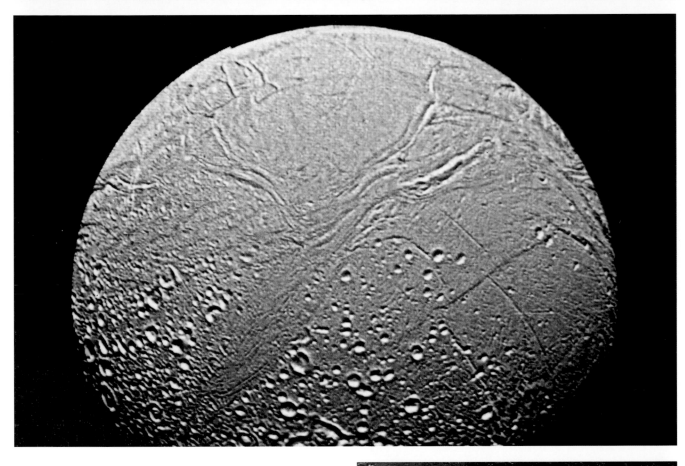

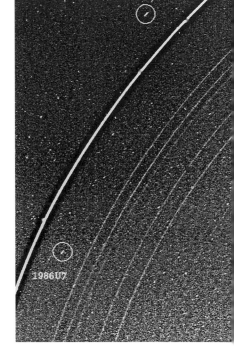

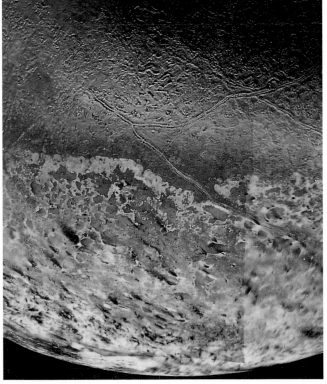

The ringed planet Saturn has even more moons than Jupiter – at least 22. The largest is Titan, unique among moons because it has a thick atmosphere of nitrogen and methane. It may have methane lakes on its surface. Among the new moons discovered by the Voyager probes are so-called "shepherd" moons. They orbit near the rings, and their gravity helps to keep the ring particles in place.

We can see only five of Uranus's 15 or so moons through a telescope. Of these, Miranda (diameter 300 miles, 485 kilometers) has the most extraordinary surface, which is a patchwork of different geological features, with sharp boundaries between them. For example, rolling plains suddenly give way to mysterious grooved regions. This seems to suggest that long ago Miranda collided with a large asteroid and was shattered to pieces. Then the pieces recombined under gravity to create the weird landscape we now see.

We can see only two of Neptune's eight or more moons from Earth – Triton and Nereid. Triton (diameter about 1,680 miles, 2,700 kilometers) was photographed closely by Voyager 2 and proved to be a pinkish world, peppered by icy "geysers." It is the coldest place we know in the solar system, with a temperature of −393°F (−236°C).

◄ Fault lines and craters scar the icy surface of Saturn's moon Enceladus (300 miles, 500 kilometers in diameter). This moon is one of the most reflective bodies in the solar system: it reflects light even better than fresh snow!

▲ Two of the tiny new moons of Uranus discovered by Voyager 2 in 1986. They are "shepherd" moons on either side of the Epsilon ring, helping to keep the ring particles in place.

◄ Close-up of the southern hemisphere of Neptune's moon Triton. It has a polar cap of pinkish snow, probably made of frozen methane and nitrogen gases. The dark plumes are probably material thrown out by erupting volcanoes or geysers.

◄ Telescope domes at the Roque de Los Muchachos Observatory on La Palma in the Canary Islands. The telescopes are, from the left, the William Herschel, the Isaac Newton, and the Jacobus Kapteyn. The Observatory occupies a superb mountain-top site at an altitude of some 7,900 feet (2,400 meters).

7

ASTRONOMERS AT WORK

When darkness falls and most of the working population head for home, supper and bed, astronomers open up the domes of their telescopes and train them on distant worlds in the unfathomable depths of space. Perhaps tonight they will search for that elusive tenth planet which some astronomers are convinced lies beyond Pluto. Or they will chart the course of a maverick, long-tailed comet. Or if they are very lucky, they will chance upon one of the most spectacular and most cataclysmic events in the Universe, a supernova.

Amateurs, too, though they have more modest instruments, still help push back the frontiers of astronomy. The Universe is a big place, and even the professionals can examine only small bits of it at any time. The huge army of amateurs dotted around the world, who each night keep sky-watch, perform inestimable service in recording the vagaries of variable stars, discovering comets newly emerged from the Stygian gloom of the outer solar system, and spotting outbursts by novae and occasionally supernovae.

Astronomy, as the T-shirt slogan goes, is looking up.

Observing the Heavens

High up on mountaintops – and in suburban backyards – astronomers pursue their career or hobby that takes them, in the mind at least, out of this world.

▼ Telescope domes at Kitt Peak National Observatory, near Tucson, in Arizona. The dome on the right houses the Mayall 157-inch (4-meter) reflector, the largest instrument on site. Kitt Peak is one of the world's finest observatories, located in the Quinlan Mountains on a peak 6,875 feet (2,095 meters) above the Sonora Desert. The local Papago Indians call Kitt Peak astronomers the People with the Long Eyes.

Professional astronomers carry out their telescopic observations at observatories, which usually house a variety of instruments. Most leading observatories are located at high altitude on mountaintops in dry climates. There they are above the thickest part of the atmosphere, which is laden with dust and water vapor, and prone to distorting air currents.

Kitt Peak (*below*) is a fine example, being located on a mountaintop in an arid desert region. Two of the finest observatories in the Southern Hemisphere are located within about 75 miles (120 kilometers) of each other in the foothills of the Andes in the Atacama Desert, where in some parts rain has not fallen for centuries! They are the Cerro Tololo Inter-American Observatory, which has a 157-inch (4-meter) reflector, and the European Southern Observatory, which has two 141-inch (3.6-meter) reflectors.

▲ The dome on Palomar Mountain, California, that houses the Hale 200-inch (5-meter) reflector, completed in 1948. Astronomers working with the new telescope, such as Edwin Hubble, pushed back the frontiers of astronomy as never before.

◀ Stonehenge on Salisbury Plain probably functioned as an astronomical observatory 4,000 years ago. This picture shows part of the outer ring of "sarsens," lined up with the distant "heel" stone.

▲ The scattered domes at the Anglo-Australian Observatory, Siding Spring, near Coonabarabran in New South Wales, Australia. The dome on the right houses the 154-inch (3.9-meter) Anglo-Australian Telescope, one of the most powerful instruments in the Southern Hemisphere.

▼ This tall tower is one of the solar telescopes at Mount Wilson Observatory, near Los Angeles, California. The Observatory is also the home of the 100-inch (2.5-meter) Hooker telescope, which first revealed the presence of external galaxies.

The best-known telescope, however, is the original giant, the 200-inch (5-meter) Hale Telescope at Palomar Observatory near Los Angeles. Alas, this historic giant is now suffering badly from something that is beginning to affect astronomers increasingly – light pollution.

Such large instruments are prodigious light-gatherers. It is said they are able to detect the light from a candle tens of thousands of miles away! But no one looks through them any more. They are used as massive cameras to record images on photographic film. Film in effect stores the light falling on it and in this way is able to detect very faint stars and galaxies.

Amateur astronomers on the other hand invariably do look through their telescopes, and some build observatories to house them. A simple design for an observatory is the so-called run-off shed. This is essentially a small shed split into two sections that can be moved apart when the telescope is required. However, it offers the observer no protection from the elements.

Such protection is essential when observing, which by its nature is a fairly static occupation. So wrap up well. In winter, thick sweater, lined jacket, two pairs of socks, thick boots, and a wool hat are to be recommended. So are a hot drink and a snack!

Among the other equipment you will need is a flashlight with a red bulb or filter so that it gives out a dim red light. Red light affects your night vision the least. You will need a flashlight to look at your equipment and another essential item, a good star map.

For your own records, if not for posterity, you should make notes on your observations on a pad, recording:

(1) The date and the time – use the 24-hour clock and follow Universal Time (Greenwich Mean Time), which means allowing for daylight savings time when it is in force.

(2) The "seeing," or viewing conditions, using the Antoniadi scale of I (ideal viewing conditions) to V (very poor conditions).

(3) Details of the telescope or binoculars used, the aperture, and the magnification.

▲ America's Stonehenge, located near Salisbury, New Hampshire, also has alignments that mark the positions of the rising and setting Sun over the seasons as they would have been in about 1500 B.C.

◀ Photographing circular star trails could not be simpler. Fix your camera on a tripod, point it in the direction of the celestial pole, open the shutter on the "B" setting (time exposure), and let the Earth rotate!

▶ You can take interesting star-trail pictures by pointing the camera at the horizon. Long-exposure photographs will then show up not only the star trails, but also a silhouette of the skyline.

▶ (*Bottom*) To photograph the stars as points rather than trails, you need a motor drive to move the telescope/camera to follow the rotation of the heavens. This picture of the Pleiades was captured with a five-minute exposure on Tri-X film.

▲ For photographing solar eclipses, as here in Hawaii at the July 1991 eclipse, ordinary cameras are fine. But you must fit a mylar aluminized filter to photograph the partial phases. The filter cuts out most of the Sun's otherwise blinding light. You remove the filter during totality, but must put it in place immediately after photographing the "diamond ring."

PHOTOGRAPHING THE HEAVENS

If you have a camera, you can photograph the night sky and begin astrophotography. Even with the simplest equipment, you can snap the Moon, stars, planets, meteors, asteroids, comets, and eclipses.

Astrophotography combines what are almost certainly the two most widely practiced hobbies in the world – astronomy and photography. Once you have taken your first rolls of night-sky pictures, you are likely to become addicted – as long as you do not expect at first to capture the superb color images that adorn the covers of astronomy magazines.

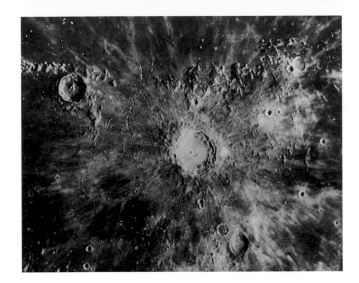

▲ Copernicus and its surroundings, photographed through a telescope. However, you can take good photographs of the Moon with a tripod-mounted camera and telephoto lens.

▼ An attractive pair of nebulae to photograph using a driven telescope: M16 (left) in Serpens Cauda and M17 in Sagittarius, both about the seventh magnitude.

▲ The nearby galaxy M33 in Triangulum is worth studying with driven telescope and camera. Photographs will begin to reveal its wide-open spiral structure. M33 is just on the limit of naked-eye visibility. It lies at about the same distance as the Great Spiral in Andromeda and is located quite near it in the sky.

Volumes have been written about astrophotography, and you are advised to read one if you want to take it up seriously. Here we deal with a few basics that will help get you started. The first essential is a camera with a facility for time exposure (often called a "B" setting). This is because the light from the stars is feeble, and the shutter needs to be left open for a while so that the film can "store" the light that falls upon it. You will not, however, need a time exposure for the Moon, because it is so bright.

The second essential is a sturdy tripod. This is needed to hold the camera absolutely rigid during the time exposure. If the camera wobbles, you will get a blurred image. Hanging extra weights on the tripod is also a good idea.

A wide selection of photographic films with different speeds is available. If you want to record stars as points in your pictures, you must use a fast film. This is because in less than half a minute, the stars will start to "trail" because of the rotation of the sky.

Good results can be obtained using fast film with an ISO (ASA) rating of 400, especially when it is "pushed," or developed as if it were a much faster film. Very fast films indeed are now available, with ISO (ASA) ratings of 1600 or more, although they tend to produce pictures that are "grainy." If you want simply to produce star trails, you can use much slower films (ISO 100 and 200), and you will be rewarded by finer detail and better color.

To improve your stellar photography further, you need to drive your camera to follow the rotation of the heavens. This will enable you to take long time exposures, which will begin to record the splendor of deep sky objects, such as nebulae, clusters, and galaxies. If you have a telescope with a motor drive, you can mount the camera on it, or you can set the camera on the telescope and use the telescope as a supertelephoto lens.

▶ The salient features of a Newtonian reflector, the type most favored by amateur astronomers. The eyepiece is on the side and at a convenient height for viewing.

Finderscope

Eyepiece

Body tube

Cradle

Polar axis

Declination axis

Counterweight

Sturdy tripod

▼ Explaining the principles of the telescope to a budding astronomer. This model is a 8½-inch (21-centimeter) Newtonian reflector. This is a good size for the dedicated stargazer.

▼ (*Right*) A well-made 4-inch (10-centimeter) refractor like this can produce excellent images. Note the finderscope and the knob at the viewing end, which is used for focusing. The eyepiece is carried by a right-angled prism mounted in the eyepiece tube. This makes for more comfortable viewing when studying stars at high elevations.

TELESCOPES

We can see much in the night sky just with the naked eye, but to appreciate its splendor fully we need to use binoculars or a telescope. These instruments have many times the light-gathering power of the eye.

There are two main kinds of telescopes – refractors and reflectors. Both produce an upside-down image, but this does not matter for astronomical work.

Refractors use glass lenses to gather and focus starlight. They have at the far end an objective lens, or object glass. This gathers the starlight and brings it to a focus. The image is then viewed and magnified by an eyepiece lens. Binoculars are in effect a pair of refractors, one for each eye.

Reflectors use mirrors to gather starlight. There are a number of different types. The one favored by most amateurs is the Newtonian, named after Isaac Newton, who built the first reflector in about 1668. It has a curved primary mirror, which gathers the light and reflects it on to a plane (flat) secondary mirror, which in turn reflects it into an eyepiece. The eyepiece is located near the top of the telescope tube.

The Cassegrain is another type of reflector. The light gathered by the curved primary mirror is reflected by a

▲ Polishing the 165-inch (4.2-meter) mirror of the William Herschel Telescope, now installed at the Roque de los Muchachos Observatory on La Palma, in the Canary Islands. It is one of the world's finest telescopes.

▼ The Anglo-Australian Telescope at Siding Spring, New South Wales, which has a 154-inch (3.9-meter) mirror. The open body tube is mounted inside a huge "horseshoe" bearing. The prime-focus cage is positioned nearly 43 feet (13 meters) above the mirror.

curved secondary mirror back through a hole in the primary and viewed there by an eyepiece.

Choosing binoculars

Binoculars are described by their power of magnification and the aperture, or diameter of their object glass in millimetres. Useful sizes for ordinary astronomical viewing are 8 × 40, 7 × 50, and 10 × 50. 10 × 50s are quite bulky to hand-hold steadily and are best mounted on a tripod for observing, as should binoculars of higher magnification and aperture to prevent image-shake.

Choosing a telescope

There are good telescopes and there are cheap telescopes; there are few, if any, good cheap telescopes! The cheaper the telescope, the poorer its optics, and so the poorer the image. Before you buy, you would be well advised to seek advice from an experienced stargazer in one of the many astronomy groups or clubs found across the country.

About the smallest useful size for an astronomical refractor is about 2½–3 inches (60–75 millimeters) aperture, or about double this for a reflector. Refractors are in general more robust than reflectors and need less attention, but are comparatively expensive.

However good the optics of your telescope, you will get nothing but poor images if it is badly mounted. Most telescopes are mounted on a tripod, which must be solidly made so that it stays absolutely firm when you are focusing or rotating the telescope.

The way the telescope is mounted on the tripod or other support is also important. The simplest mounting is the altazimuth. This allows you to angle the telescope up and down and also horizontally. But this mounting makes it awkward to follow the path of the stars, which travel neither vertically nor horizontally.

The best mounting is the equatorial. This allows the telescope to move around an axis (the polar axis) which is parallel with the Earth's axis and points toward the celestial pole. The path of the stars, which circle around the celestial pole, can then be followed simply by rotating the telescope around the polar axis.

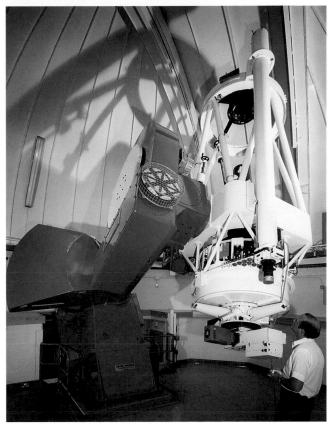

▲ The 39-inch (1-meter) Jacobus Kapteyn reflector at the Roque de los Muchachos Observatory on La Palma, in the Canary Islands.

▼ The CCD (charge-coupled device) is revolutionizing optical astronomy. It is very much more sensitive to light than photographic film.

RADIO AND SPACE TELESCOPES

▲ The Arecibo radio telescope in Puerto Rico, the biggest in the world, consists of a dish 1,000 feet (305 meters) across. The dish is made of 38,778 perforated aluminum panels suspended in a natural bowl in a mountain-top. It collects radio waves from the heavens and focuses them on the antenna suspended high above.

▲ (*Right*) This huge tank of dry-cleaning fluid deep in a mine in South Dakota, must be the oddest astronomical "instrument." It acts as a detector for neutrinos, which are produced in the heart of the Sun.

▶ The largest radio telescope of the Australian National Radio Observatory at Parkes, New South Wales, has a dish 210 feet (64 meters) across.

Some of the most dramatic astronomical discoveries of recent years, such as quasars and pulsars, have been made by using radio telescopes to study the radio waves stars and galaxies give out. Telescopes sent into space to pick up both visible and invisible wavelengths have also made exciting new discoveries: of stars being born, of other solar systems, and of possible black holes.

Stars give out their energy not only as visible light rays, but also as invisible gamma rays, X-rays, ultraviolet rays, infrared rays, microwaves, and radio waves. All these rays are different kinds of electromagnetic radiation which differ from one another in their wavelengths. Gamma rays have the shortest wavelengths, and radio waves the longest.

The stars do not give out energy equally at all wavelengths. For example, a star that looks dim in visible light may shine like a beacon when viewed, say, at radio wavelengths. So to get a true picture of the heavens, we must view them at all wavelengths.

The trouble is that the Earth's atmosphere blocks most invisible wavelengths. Fortunately, it does let through radio waves, which makes possible radio astronomy from the ground. This science has its origins in 1931 when Karl Jansky, an American Bell Telephone Laboratories researcher, discovered radio waves coming from the heavens. But radio astronomy did not really take off until after the construction of giant radio telescopes, such as the 250-foot (76-meter) dish at Jodrell Bank near Manchester, England, which was completed in 1957.

The Parkes radio telescope in Australia (left) is another powerful instrument. The dish gathers radio waves from the heavens and concentrates them on the antenna above. The signals are then fed through electronic circuits to a computer, which can display a visible image of the radio source (see page 154).

Space telescopes

The atmosphere blocks the gamma rays, X-rays, ultraviolet rays and infrared rays that reach the Earth from the stars. So astronomers have to send instruments into space on satellites to detect them. The results have been spectacular and have given astronomers quite a different view of the universe.

Among the most outstanding satellites have been the X-ray satellites Einstein and Exosat, IUE (International Ultraviolet Explorer), IRAS (Infrared Astronomy Satellite), and the Compton (Gamma Ray) Observatory.

Optical telescopes, too, are carried into orbit, where they can take advantage of a position above the Earth's obscuring atmosphere. The space shuttle regularly carries astronomy payloads, as does the Russian space station Mir. But the outstanding instrument is the Hubble Space Telescope, which is producing good images despite a manufacturing fault.

▲ The telescope cluster on Astro-1, an astronomy payload carried into space in December 1990 by space shuttle Columbia. It looked at the Universe at ultraviolet wavelengths, which do not reach Earth because they are absorbed by the atmosphere.

▶ (*Top*) Two Voyager probes of this design, launched in 1977, have revolutionized the study of the outer planets. Their instruments include cameras, radiometers, and particle detectors. For scale: the dish antenna measures 144 inches (3.7 meters) across.

▶ The Hubble Space Telescope, pictured in orbit. This 12-ton (11-tonne) satellite, which operates at optical wavelengths, has a mirror 95 inches (2.4 meters) across. Despite faulty optics, it has been returning better images than Earth-bound telescopes.

INVISIBLE IMAGES

Most of the pictures in this book are photographs of heavenly bodies taken with ordinary light, that is, at visible wavelengths. They show the bodies as we would see them with our eyes. The images on these pages, however, are pictures taken at invisible wavelengths. They show what we would see if our eyes were sensitive to these wavelengths.

The images have been produced from the signals picked up by detectors sensitive to the particular wavelengths under study. Computers have processed the signals and arbitrarily assigned different colors to different intensities of radiation. This results in so-called false-color images.

The images have been selected so that they cover different wavelengths of the electromagnetic spectrum, from the short-wavelength gamma rays to the long-wavelength radio waves.

▶ X-ray
An image of the cluster of galaxies called Abell 2256, returned by the German X-ray satellite Rosat (Roentgen satellite).

▼ Ultraviolet
The ultraviolet imaging telescope of the shuttle-borne satellite Astro-1 returned this image of the face-on spiral galaxy M74 in Pisces. The bright points highlight regions of intense star formation.

▼ Infrared
Comet IRAS-Araki-Alcock of 1983, first detected by IRAS (infrared astronomy satellite). During its 300-day mission, IRAS also peered into the center of our Galaxy; spotted stars being born in nebulae; and detected disks of matter around other stars, which may eventually condense into planets.

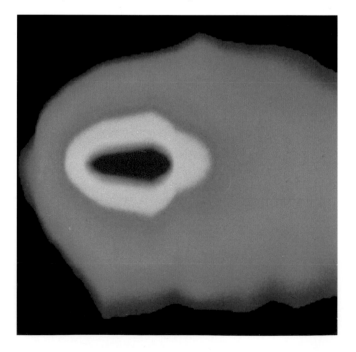

▶ Microwave

A microwave map of the whole sky, produced from data returned by COBE (cosmic background explorer). The color differences reflect very slight variations in the background temperature, which averages about three degrees above absolute zero. Such variations are predicted by the Big Bang theory of the creation of the universe.

▼ Microwave radar

The planet Venus, as we could never see it in visible light. This global view is a mosaic of images obtained by radar scans by the orbiting Magellan space probe in 1990/91.

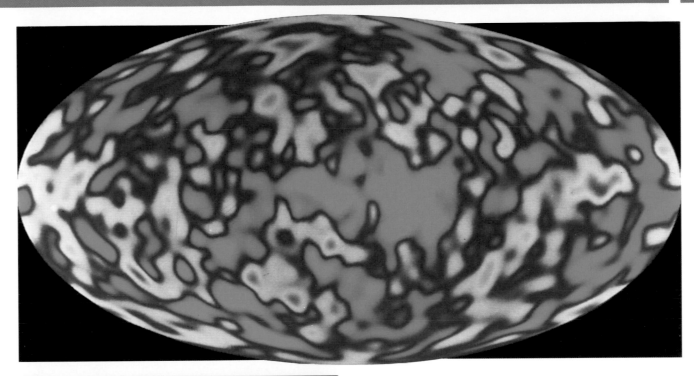

▼ Radio

A false-color image of the radio source 3C75 in the Abell 400 cluster of galaxies in Cetus. NRAO's Very Large Array radio telescope near Socorro, New Mexico, gathered the data using its 27 antennae. The image shows twin radio "jets" coming from the nucleus of the central galaxy in the cluster.

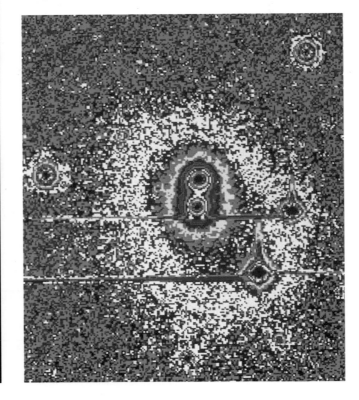

Glossary

Absolute magnitude   A measure of the true brightness of a star. It is what the brightness of the star would be if viewed from a distance of 10 parsecs, or 32.6 light-years.

Active galaxy   One that has an exceptionally high energy output.

Antoniadi scale   A scale used by an observer to indicate the quality of seeing.
I   Perfect seeing
II   Mainly calm, but slight disturbance
III   Moderate seeing, more disturbance
IV   Poor seeing, constant disturbance
V   Very bad seeing, very bad disturbance

Apparent magnitude   A measure of the brightness of a star as we see it from Earth.

Asteroids   Small rocky bodies that circle the Sun, mainly in a "belt" between the orbits of Mars and Jupiter. Also called the minor planets and planetoids.

Astrology   Study of the positions of the heavenly bodies with a view to explaining events in people's lives or foretelling the future.

Astronomical unit (AU)   The distance between the Earth and the Sun, 93,000,000 miles (149,600,000 kilometers).

Astronomy   The scientific study of the universe and the bodies therein – the Sun, the Moon, the planets, their moons, the stars, and the galaxies.

Aurora   A colorful glow seen mainly in far northern and far southern skies, produced when charged particles from the Sun collide with particles in the upper atmosphere. In northern skies it is called the aurora borealis, or Northern Lights; and in southern skies the aurora australis, or Southern Lights.

Big Bang   The name for the event that is believed to have created the universe, some 15,000 million years ago.

Binary   A two-star system in which the stars are associated and circle around a common center of gravity (barycenter).

Black hole   A region of space in which gravity is so immensely strong that not even light can escape from it.

Celestial sphere   An imaginary sphere around the Earth, to the inside of which the stars seem to be fixed.

Cepheids   A common class of variable stars that vary in brightness as regularly as clockwork. They are named after the prototype star Delta Cephei.

Comet   A small member of the Solar System, a "dirty snowball" of rock, ice, and dust, which starts to shine when it approaches the Sun.

Conjunction   The lining up of heavenly bodies in the sky, such as a planet, the Sun, and the Earth.

Constellations   Imaginary patterns that the bright stars make in the night sky.

Cosmology   The study of the origin, evolution, and structure of the universe.

Culmination   The maximum altitude of a heavenly body above the horizon; this happens when it crosses the meridian.

Declination   A star's celestial latitude on the celestial sphere, measured in degrees north (+) or south (−) of the celestial equator.

Double star   A pair of stars that appear close together in the sky. In a binary star, the two components are physically associated. In an optical double, the components are usually quite separate, appearing together only because they lie in the same direction.

Eclipse   The passing of one heavenly body in front of another, blotting out its light. A solar eclipse, or an eclipse of the Sun, occurs on Earth when the Moon passes in front of the Sun. A lunar eclipse, or an eclipse of the Moon, occurs when the Moon moves into the Earth's shadow in space.

Eclipsing binary   A binary-star system that varies in brightness when its two components periodically pass in front of one another in our line of sight.

Ecliptic   The apparent path of the Sun through the heavens during the year.

Elongation   The angular distance between a planet and the Sun.

Equinoxes   Times of the year when the lengths of the day and the night are equal all over the world because the Sun is directly over the Equator. This happens twice: on March 21 (the vernal, or spring equinox); and on September 23 (the fall equinox).

Evening star   A planet (usually Venus) that shines brightly in the western sky at sunset.

Expanding Universe   The theory that the universe is getting bigger. Astronomers believe that expansion began with the Big Bang.

Fireball   An exceptionally bright meteor; also called a bolide.

Galaxy   A star "island" in space. Galaxies are usually elliptical or spiral in shape. We usually call our own galaxy, the Galaxy, or the Milky Way.

Globular cluster   A globe-shaped group of stars numbering up to a million or more.

Gravity   The force with which the Earth attracts any object near it. The other heavenly bodies exert a similar force. Gravity is the force that literally holds the universe together.

Librations   Oscillations of the Moon as we see it from Earth that enable us to see slightly "around the edges" of the face it presents.

Light-year   A common unit for measuring distances in space. It is the distance light travels in a year: 5.88 million million miles (9.46 million million kilometers).

Limb   The edge of the visible disk of a planet or a moon.

Luminosity   A measure of the total energy given off by a star.

Lunar   Relating to the Moon.

Magnitude   The scale on which star brightness is measured. On the scale, the visible stars are divided into six levels of brightness, and the scale is extended to describe brighter and dimmer stars. See also *Absolute magnitude; Apparent magnitude.*

Mare   A large plain on the Moon. "Mare" (plural "maria") is the Latin word for "sea." The dark areas we see on the moon are maria. The main sea areas (and English names) are:
Lacus Mortis (Lake of Death)
Lacus Somniorum (Lake of Dreams)
Mare Crisium (Sea of Crises)
Mare Fecunditatis (Sea of Fertility)
Mare Frigoris (Sea of Cold)
Mare Humorum (Sea of Humors)
Mare Imbrium (Sea of Showers)
Mare Nectaris (Sea of Nectar)
Mare Nubium (Sea of Clouds)
Mare Orientale (Eastern Sea)
Mare Serenitatis (Sea of Serenity)
Mare Smythii (Smyth's Sea)
Mare Tranquillitatis (Sea of Tranquillity)
Mare Vaporum (Sea of Vapors)
Oceanus Procellarum (Ocean of Storms)
Sinus Iridum (Bay of Rainbows)
Sinus Medii (Central Bay)
Sinus Roris (Bay of Dews)

Meridian   The great circle on the celestial sphere that passes through the north and south poles. An observer's meridian passes through the north and south points on the horizon and the zenith.

Meteor   A streak of light we see when a piece of rock from outer space plunges through the atmosphere and burns up.

Meteorite   A piece of rock from outer space big enough to survive its passage through the atmosphere and reach the ground.

Milky Way   A pale band of light that we can often see arching across the heavens. It is a region dense with stars and is, in effect, a cross-section of our Galaxy, which is also called the Milky Way.

Moon   A natural satellite of a planet.

Morning star   A planet (usually Venus) that shines brightly in the eastern sky at dawn.

Nadir   The point on the celestial sphere directly beneath an observer.

Nebula   A cloud of gas and dust between the stars. We see bright nebulae because they glow or reflect light; we see dark nebulae because they obscure light from distant stars.

Neutron star   An incredibly dense body made up of a solid mass of neutrons. See also *Pulsar*.

Nova   A "new" star; actually one that suddenly increases greatly in brightness.

Nuclear fusion   The joining together of the nuclei (centers) of light atoms, such as hydrogen, to make heavier ones, such as helium. It is the energy-production process in stars.

Occultation   The temporary hiding of one heavenly body by another, for example of a star or a planet by the Moon.

Open cluster   A loose grouping of stars, typically containing a few hundred members.

Opposition   The position of a planet in its orbit when it lies exactly opposite the Sun in the sky.

Parallax principle   The apparent change in position of a nearby object against a distant background when looked at from different angles. The parallax of a star is the angle subtended at the star by the Earth's radius.

Parsec   The unit professionals generally use to measure distances in space. It is the distance at which the parallax of a star would be one second of arc (1/3600th of a degree). Equal to 3.26 light-years.

Phases   The different shapes of the Moon we see in the sky during the month, as more or less of its surface is lit up by the Sun. Venus shows noticeable phases as well.

Planets   Large bodies that circle in space around the Sun. Some of the other stars probably have planets circling around them.

Precession   A slow change in the direction in which the Earth's axis points in space, which over the years alters the positions of the stars on the celestial sphere.

Proper motion   Observed motion of a nearby star against the stellar background.

Pulsar   A rapidly rotating neutron star that gives off pulses of radiation.

Quasar   A remote body that looks like a star but has the energy output of many galaxies. Also called quasi-stellar object (QSO).

Radial motion   Motion of a star toward or away from us, detectable from its spectrum.

Retrograde motion   Motion of a heavenly body in the opposite direction from usual.

Right ascension   A star's celestial longitude, expressed usually in hours and minutes of sidereal time. It is measured from the point (First Point in Aries) where the ecliptic meets the celestial equator at the vernal equinox.

Saros   A period of a little over 18 years (6,585.3 days), after which the Moon returns to exactly the same position in relation to the Sun and the Earth. It therefore marks the interval between successive eclipses of the same type.

Satellite   A small body that circles around a larger one in space; a moon. Most of the planets have natural satellites. The Earth has one, the Moon. Saturn has more than 20. Earth also now has thousands of artificial satellites circling around it, launched into orbit by space scientists.

Seasons   Periods of the year marked by noticeable changes in the weather, especially in temperature. They occur because of the tilt in the Earth's axis with respect to the plane of its orbit around the Sun.

Seeing   The quality of the observing conditions at the time of observation.

Sidereal time   Time measured in relation to the stars, based on the sidereal day, the Earth's true period of rotation on its axis, 23 hours 56 minutes 4 seconds of ordinary clock time.

Solar System   The family of the Sun that travels through space as a unit. At the center is the Sun, around which circle nine planets (including the Earth), the asteroids, and numerous comets.

Solar wind   A stream of charged particles given off by the Sun.

Solstices   Times of the year when the Sun reaches its highest and lowest points in the sky at noon. In the Northern Hemisphere, it reaches its highest point on about June 21 (summer solstice), and its lowest on December 21 (winter solstice). The dates are reversed in the Southern Hemisphere.

Spectrum   A band of color (spread of light wavelengths) produced when light is passed through a spectroscope. Stellar spectra are crossed by dark lines.

Speed of light   Light travels in a vacuum at a speed of about 186,000 miles (300,000 kilometers) a second. It is the fastest speed possible.

Star   A gaseous body that produces energy by nuclear fusion. It releases this energy as light, heat, and other radiation. The Sun is our local star.

Sunspot   An area on the Sun's surface that is darker and cooler than normal. Sunspots come and go regularly over a period of about 11 years.

Supernova   A gigantic stellar explosion in which a supergiant star blasts itself apart.

Supernova remnant   An expanding shell of gas resulting from a supernova explosion.

Terminator   The boundary between the sunlit and dark halves of the Moon or a planet.

Transit   The movement of one small heavenly body in front of a larger one, such as Venus across the disk of the Sun.

Universe   Everything that exists: the Earth, the Moon, the Sun, the planets, the stars, and even space itself.

Variable star   One that changes in brightness. Intrinsic variables (such as Cepheids) vary in brightness because of internal physical changes. Extrinsic variables (such as eclipsing binaries) vary because of some external influence.

Zenith   The point on the celestial sphere directly above an observer.

Zodiac   An imaginary band in the heavens in which the Sun, Moon, and planets are always found. It is occupied by 12 constellations, the constellations of the zodiac.

INDEX

Constellations are shown in capitals. Page numbers in bold refer to main entries. Page numbers in italics refer to illustrations or maps. Textual matter may also occur on the same pages.

Deep sky objects identified only by M or NGC numbers which appear on maps but not in the text are omitted from the index as are lunar features which appear only in the quadrants.

ACKNOWLEDGMENTS

The author and publishers would like to thank Spacecharts Photo Library for picture research and for supplying many of the photographs in the book. They are also indebted to the following establishments for providing invaluable information and illustrative material: Anglo-Australian Observatory, Astronaut Memorial Planetarium and Observatory, European Space Agency, Goddard Space Center, Jet Propulsion Laboratory, Kitt Peak National Observatory, London Planetarium (p.35), National Aeronautics and Space Administration, National Radio Astronomy Observatories, Palomar Observatory, Parkes Radio Observatory, Roque de los Muchachos Observatory, Royal Astronomical Society, Royal Greenwich Observatory, US Naval Observatory.

Photographs on the following pages were taken by the author: 6, 90, 102t, 126tr, 128, 129, 134bl, 136b, 146, 147, 148t, 150bl, 151, 152lt, 152b. The photograph on page 73 was taken by David Malin, courtesy Anglo-Australian Observatory.

While every effort has been made to trace and acknowledge all copyright holders, we would like to apologize should there have been any omissions.

The publishers would also like to thank Moira Greenhalgh for supplying the index.